Fundamental and Applied Problems of Mechanics (FAPM 2020)

IOP Conference Series: Materials Science and Engineering
Volume 1191

Moscow, Russia
2 – 4 December 2020

ISBN: 978-1-7138-4651-2
ISSN: 1757-8981

Printed from e-media with permission by:

Curran Associates, Inc.
57 Morehouse Lane
Red Hook, NY 12571

Some format issues inherent in the e-media version may also appear in this print version.

Institute of Physics
Dirac House, Temple Back
Bristol BS1 6BE UK

Phone: 44 1 17 929 7481
Fax: 44 1 17 920 0979

techtracking@iop.org

Additional copies of this publication are available from:

Curran Associates, Inc.
57 Morehouse Lane
Red Hook, NY 12571 USA
Phone: 845-758-0400
Fax: 845-758-2633
Email: curran@proceedings.com
Web: www.proceedings.com

TABLE OF CONTENTS

PREFACE

The conference held in the Bauman Moscow State Technical University. The main purpose of this conference is to present new research results in various areas of mechanics held in Russian and foreign scientific and educational institutions. The conference was attended by Dr. Christian Bizouard (Director of IERS Earth Orientation Centre, Observatoire de Paris), Prof. Dr. Shuanggen Jin (Head of Satellite Navigation and Remote Sensing Group, Chinese Academy of Sciences) and other famous scientists. The scientists showed their great potential in studying various problems in numerous fields of science and technology, as well as applying new approaches, which develop and improve methods of mechanics. The reports demonstrated that differential equations and theoretical methods of dynamics present a great potential in describing the motion of the Earth's pole, the approach of the Earth and the asteroid, in problems of gravitational-wave astronomy, and also in other classical problems of mechanics.
Since 2017, Conference Materials have been successfully published by IOP Publishing.

International Organizing Committee of the FAPM-2020:

- Professor A.A. Aleksandrov (Chairman), Bauman Moscow State Technical University Moscow, Russia, *rector@bmstu.ru*
- **Professor P.M. Shkapov (Co-Chairman), Bauman Moscow State Technical University Moscow, Russia, *spm@bmstu.ru***
- **Astronom C. Bizouard (Co-Chairman), Observatoire de Paris, France, *christian.bizouard@obspm.fr***
- Professor V.O. Gladyshev, Bauman Moscow State Technical University, Moscow, Russia, *dekan-fn@mail.ru*
- Professor K. Matsumoto, RISE Project, National Astronomical Observatory of Japan, Japan, *koji.matsumoto@nao.ac.jp*
- Professor V.S. Aslanov, Samara University, Samara, Russia, *aslanov_vs@mail.ru*
- Professor B.S. Bardin, Moscow Aviation Institute (National Research University), Moscow, Russia, *bsbardin@yandex.ru*
- Professor V.D. Bertyaev, Tula State University, Tula, Russia, *tpi-vit-59@yandex.ru*
- Professor S.A. Berestova, Ural Federal University, Yekaterinburg, Russia, *s.a.berestova@urfu.ru*
- Professor I.G. Blagoveshchenskiy, Bauman Moscow State Technical University, Moscow, Russia, *drbl@bk.ru*
- Professor Yu.I. Dimitrienko, Bauman Moscow State Technical University, Moscow, Russia, *dimit.bmstu@gmail.com*
- Professor V.V. Ivashkin, Keldysh Institute of Applied Mathematics, Russian Academy of

IOP Conf. Series: Materials Science and Engineering 1191 (2021) 011001 doi:10.1088/1757-899X/1191/1/011001

Sciences, Moscow, Russia, *ivashkin@keldysh.ru*

- Professor V.T. Kalugin, Bauman Moscow State Technical University, Moscow, Russia, *kaluginvt@mail.ru*
- Professor V.A. Kalinichenko, Ishlinsky Institute for Problems in Mechanics, Russian Academy of Sciences, Moscow, Russia, *vakalin@mail.ru*
- Professor G.N. Kuvyrkin, Bauman Moscow State Technical University, Moscow, Russia, *fn2@bmstu.ru*
- Professor V.I. Mayorova, Bauman Moscow State Technical University Moscow, Russia, *victoria.mayorova@gmail.com*
- Professor A.N. Morozov, Bauman Moscow State Technical University Moscow, Russia, *amor@bmstu.ru*
- Professor M.Yu. Ovchinnikov, Keldysh Institute of Applied Mathematics, Russian Academy of Sciences, Moscow, Russia, *ovchinni@keldysh.ru*
- Professor S.N. Sayapin, Blagonravov Mechanical Engineering Research Institute, Russian Academy of Sciences, Moscow, Russia, *s.sayapin@rambler.ru*
- Professor Yu.P. Ulybyshev, S. P. Korolev Rocket and Space Corporation Energia, Korolyov, Russia, *yuri.ulybyshev@rsce.ru*
- Professor V.N. Chubarikov, Lomonosov Moscow State University , Moscow, Russia, *chubarik2009@live.ru*
- Professor A.O. Shimanovskiy, Belarusian State University of Transport, Gomel, the Republic of Belarus, *tm.belsut@gmail.com*
- **Academic Secretary - Ph.D. E.V. Melkumova, *elena_v_m@mail.ru***

Academician Committee of FAPM-2020 (International Advisory Council)

- Academician I.B. Fedorov (Chairman), Bauman Moscow State Technical University, Moscow, Russia, *president@bmstu.ru*
- **Professor V.N. Zimin (Co-Chairman), Bauman Moscow State Technical University, Moscow, Russia, *zimin@bmstu.ru***
- Professor V.O. Gladyshev (Co-Chairman), Bauman Moscow State Technical University, Moscow, Russia, *dekan-fn@mail.ru*
- **Professor H. Hanada (Co-Chairman), director of RISE Project, NAO of Japan, Japan, *hideohanadah@gmail.com***
- Astronom C. Bizouard (Co-Chairman), Observatoire de Paris, France, *christian.bizouard@obspm.fr*
- Professor Shuanggen Jin, Chinese Academy of Sciences, China, *sgjin@shao.ac.cn*
- Academician V.F. Zhuravlev, Ishlinsky Institute for Problems in Mechanics, Russian Academy of Sciences, Moscow, Russia, *zhurav@ipmnet.ru*
- Academician S.T. Surzhikov, Ishlinsky Institute for Problems in Mechanics, Russian Academy of Sciences, Moscow, Russia, *surg@ipmnet.ru*
- Academician D.V. Treshchev, Lomonosov Moscow State University, Moscow, Russia, *treschev@mi-ras.ru*

IOP Conf. Series: Materials Science and Engineering 1191 (2021) 011001 doi:10.1088/1757-899X/1191/1/011001

IOP Conf. Series: Materials Science and Engineering 1191 (2021) 011002 doi:10.1088/1757-899X/1191/1/011002

Peer review declaration

All papers published in this volume of IOP Conference Series: Materials Science and Engineering have been peer reviewed through processes administered by the proceedings Editors. Reviews were conducted by expert referees to the professional and scientific standards expected of a proceedings journal published by IOP Publishing.

- **Type of peer review:**
  Open

- **Conference submission management system:**
  Abstracts (one page of text) and Papers in English offered for inclusion in the Moscow FAPM-2020 Program should be sent to the Local Organizing Committee.

  The format of files of abstracts and papers is *.docx and *.doc (you also should send the copy in PDF format). An article should be formed as one file and should include:

  1. A title;

  2. Information about authors;

  3. Key words;

  4. A brief annotation;

  5. Text of paper;

  6. A list of references;

  7. Illustrations and tables should be numbered and titled and be included into the text of paper.

  All papers and abstracts should be formed due to the rules which you can find on the Moscow FAPM-2020 web-site.
  All enquiries concerning length of abstracts, papers, registration, accommodation, etc. should be sent to the Organizing Committee preferably by e-mail to FAPM2020@mail.ru
  The articles should be thoroughly revised and put into shape in according to requirements.
  The Proceedings will generally revised, therefore, the Organizing Committee keeps the rights for itself to use the special review of any article. In the case reviewers will be specialists of Leading Universities of Russia or aboard.

- **Number of submissions received:** 117
- **Number of submissions sent for review:** 82
- **Number of submissions accepted:** 40
- **Acceptance Rate (Number of Submissions Accepted / Number of Submissions Received X 100):** 34
- **Average number of reviews per paper:** 2
- **Total number of reviewers involved:** 18
- **Any additional info on review process:**

  First, we send an invitation to take part in our conference to as many potential participants as possible from Russian and foreign scientific and educational institutions.

  We send the resulting abstracts to one of the following sections:

  • Theoretical mechanics
  • Celestial Mechanics
  • Mathematical modeling of mechanical systems, stability, traffic control
  • Hydromechanics
  • Aerodynamics, ballistics and space flight control
  • Applied mechanics, theory of oscillations
  • Problems of optimization, identification and reliability of mechanical and hydromechanical systems
  • Biomechanics, robotics, mechatronics
  • Theory of Mechanisms and Mechanics of Machines
  • Computational mathematics in mechanics
  • Topical issues of natural sciences
  • Scientific-methodical and historical questions of mechanics

  Further, the participants must make an oral presentation at the conference. Each section has 1-2 supervisors from among well-known Russian or foreign scientists, who can, after hearing the report, recommend publishing an article on this work.
  Works must be original and new, not previously published anywhere. Have scientific novelty and relevance.
  If all these conditions are met, and there is also a recommendation from the supervisor of the section, then we accept the article in order to apply.

- **Contact person for queries (please include: name, affiliation, institutional email address)** *Shkapov Pavel, Bauman Moscow State Technical University,* barkin@bmstu.ru

IOP Conf. Series: Materials Science and Engineering 1191 (2021) 012001 doi:10.1088/1757-899X/1191/1/012001

Matching of models of Titan with tidal Love numbers

V A Kronrod [1], A N Dunaeva[1] and O L Kuskov[1]

[1]Vernadsky Institute of Geochemistry and Analytical Chemistry, Russian Academy of Sciences (RAS), Kosygina St., 19, Moscow, 119991 Russia

va_kronrod@mail.ru

Abstract. The restrictions on the heat flux, the structure of the water-ice shell, and the composition of the Titan ocean, which are necessary to match the calculated (model) and experimental Love numbers, are considered. Estimates are obtained for model Love numbers and maximum surface heat fluxes for the L/LL chondrite composition of the rock-iron component of Titan, taking into account the radiogenic and tidal energy, as well as the energy of ice crystallization in the ocean due to satellite cooling. The values of the tidal Love numbers k_2 of Titan were calculated for various models of the water-ice shell of the satellite, including the outer ice crust (ice Ih ± CO_2 clathrates) and the internal ocean of various compositions (pure or sea water, ammonia + dissolved salts solution, aqueous solutions of ammonia and methanol). Titan models with an outer ice crust (ice Ih + 5% CO_2 clathrates) and with an internal ocean of low salinity based on ammonia (5% NH_3) aqueous solution better than other tested models satisfy the k_2 experimental values.

1. Introduction

Several parameters measured by the Cassini-Huygens mission provide insight into the interior of Titan: the second-degree gravity field coefficients, the shape, the tidal Love number. In particular, the normalized moment of inertia of Titan I = 0.34190±0.001 was measured for the first time [1] and tidal Love numbers k_2 were obtained [2]. In a recent work [3], based on the processing of new gravitational data, the value I = 0.341 was published. Processing data on Titan's gravitational field in the course of six flyby of the satellite by the Cassini spacecraft resulted in calculations of two values of the Love numbers k_2: 0.589± 0.075 and 0.637±0.112 [2]. Additional processing of results from four other flybys of the spacecraft allowed one to obtain a refined value k_2 = 0.616 ± 0.067 [3]. Several lines of evidence (tides, rotation, and electric-field measurements) suggest the presence of an internal global ocean at depth. The problem of the structure of the water-ice shell (WIS) of the satellite – the internal (underice) ocean and the ice shell situated above it holds a special place in the physics of Titan. The determination of the size, composition, and physical properties of this system is necessary for understanding the thermal regime, origin history, and evolution of Titan. However, reliable data on the depth and composition of the supposed internal ocean, as well as on the thickness of the outer ice crust of the satellite, have not yet been obtained.

The main parameters limiting the thickness of the ice shelf and the internal ocean of Titan, as well as the elastic and density properties of its interior, are the Love numbers k_2, h_2, which characterize the satellite's response to tidal perturbations. The number k_2 reflects the change in the satellite's gravitational potential due to the tidal redistribution of masses in its interior. Love numbers k_2 are sensitive to the thickness of the ice crust and internal ocean of Titan, as well as elastic and density properties of its interior, and therefore can be used to analyze models of Titan with the likely presence of an internal ocean. The Love numbers are sensitive to the position of the liquid–solid body interface and, for this reason, are used in the analysis of Titan models with the probable presence of an internal ocean. Baland et al. (2011) [4] modelled the internal structure of Titan based on data on its average density and mass, assuming the existence of an internal ocean in the satellite, and obtained estimates of Love numbers k_2, h_2 depending on the density of the internal ocean, the thickness and density of the outer crust from ice Ih.

This paper presents the results on the estimates of the Love numbers k_2 for a wide range of Titan models. Constraints on the structure, dimensions of the outer water-ice shell of the satellite, the composition of the inner ocean and heat fluxes from the bowels of the satellite are discussed.

2. Titan Models

Three-layer models of the internal structure of partially differentiated Titan are investigated. Titan consists of the outer ice crust, below which the internal ocean is located, the rock-ice convective mantle, and the central rock-iron Fe-Si core [5-7]. A variant of Titan's H_2O shells containing liquid water and several solid phases, represented by hexagonal ice-Ih, high-pressure ices III, V, VI, some of which may also be part of the rock-ice mantle, are considered. The intermediate rock-ice mantle consists of a homogeneous mixture of high-pressure ices and rock-iron material (L/LL, CI chondrites). The central core is composed of L/LL, CI chondrites.

To determine the phase composition of the Titan's high-pressure ices, as well as to calculate the density distribution of matter in its water-ice shell and rock-ice mantle, the equations of state for chondrite matter, liquid water [8] and H_2O-ices were used. In the outer ice-Ih shell, a coarse-grained ice structure and conductive transport are assumed. In the water ocean and convective ice-rock mantle, the temperature gradient is set in accordance with the adiabatic gradient.

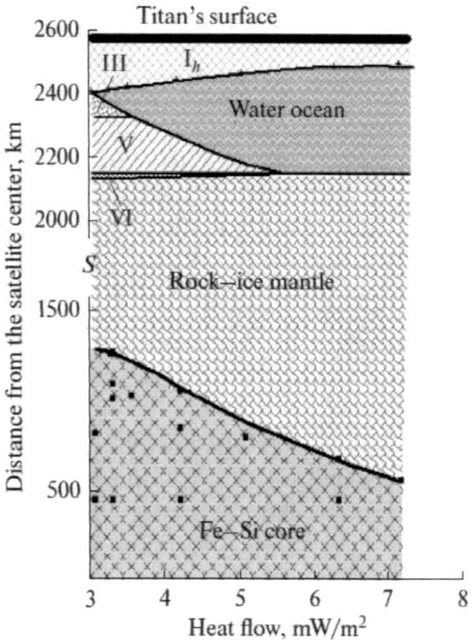

Figure 1. Models of the internal structure of partially differentiated Titan in the range of heat fluxes $3.1 < F < 7$ mW/m^2. The dots show the dimensions of the rock-iron (Fe-Si) core of the satellite at different values of the density of the rock-iron component.

Numerical modelling of the internal structure of Titan was carried out on the basis of the available geophysical information about the size, mass and reduced moment of inertia of Titan. The calculations used the equations of hydrostatic equilibrium, conservation of mass and moment of inertia of the satellite, equations of conductive heat transfer in the outer ice Ih-crust, equations of state of H_2O-phases, as well as equations for determining the concentration of the ice component in the rock-ice mantle [5, 6, 9-11].

An algorithm for calculating the radial structure of a satellite is described in [6]. The construction of the model of the rocky-ice mantle of Titan was carried out on the assumption of global convection of matter in the entire mantle reservoir and the homogeneity of the mantle composition. As a result of the solution, the following are found: distributions over the depth of the mantle of the density of high-pressure ice and their phase boundaries; mass concentration of ice in the mantle; the location of the boundary between the water-ice shell and the rock-ice mantle; the size of the rock-iron core; the water (ice)/rock ratio in Titan's mantle and in the satellite as a whole. Figure 1 shows models of the internal structure of partially differentiated Titan in the range of surface heat flow $3.1 < F < 7$ mW/m^2.

3. Calculation of model Love numbers k_2

The value of the Love numbers k_2, h_2 is mainly determined by both the size and state of the outer ice shell of the satellite, and the state of aggregation (solid or liquid) of the medium under it. To estimate the Love numbers, Titan with a sufficient degree of accuracy can be considered as an elastic inhomogeneous compressible and gravitating sphere, in which all the characteristics of its internal structure depend only on the distance to its center.

Love numbers are functionals of the density distribution ρ, the shear modulus μ, and the modulus of compression K in the planet's interior. The elastic equilibrium of a sphere is described by a coupled system of equations of the theory of elasticity and the theory of gravitational potential [12]. The shear modulus μ and the modulus of compression K are taken constant in each layer of the satellite.

Baland et al. (2014) [4] carried out massive calculations for possible models of Titan and showed that the value of the tidal Love number k_2 is more dependent on the density of the internal ocean (ρ_0), the thickness of the outer crust Ih (h_s), and to a lesser extent on the density (ρ_s) and shear modulus (μ_s) of ice Ih. An empirical dependence of the k_2 value was obtained [4]:

$$\begin{aligned} k_2 &= 0.129 + 0.286(10^{-6}\,\rho_0^{\,2}) - 0.202(10^{-8}h_s\,\rho_0) + 0.134(10^{-8}h_s\,\rho_s) - 0.097(10^{-3}\,\rho_s) \\ &+ 0.079(10^{-6}\,\rho_s\,\rho_0) + 0.078(10^{-3}\,\rho_0) + 0.039(10^{-5}h_s) + 0.017(10^{-6}\,\rho_s^{2}) - 0.012(10^{-14}\,h_s\mu_s) \\ &+ 0.010(10^{-10}h_s^{2}) + 0.010(10^{-9}\,\mu_s) - 0.010(10^{-12}\rho_0\mu_s) - 0.005(10^{-12}\rho_s\mu_s) + 0.001(10^{-18}\,\mu_s^{2}) \end{aligned} \quad (1)$$

This empirical fit reproduces most of the computed cases with an error of less than 1%, and in a few cases up to a few percents [4]. The results [4] confirm the theoretical estimates [13], where it is shown that in the case of a thin water-ice shell with respect to the satellite radius, the Love numbers k_2, h_2 are almost independent of the density and shear moduli of the underlying shells and are mainly determined by the parameters of the water-ice ice shell (density, shear moduli of ice and water; ice thickness). The depth of the ocean does not affect the values of k_2, h_2. The existence of a liquid layer with a zero or very low shear modulus between the ice crust and the rocky-ice mantle is sufficient.

4. Dependence of crustal thickness on heat flow

As shown above, to calculate the Love numbers it is necessary to know the ice thickness h_s. In most models, the thickness of the ice crust is based on the assumption of stationary conductive heat transfer in the ice layer. Boundary conditions are set on the surface of Titan: temperature $T_0 = 93$ K and heat flux $F = k_{kond}(T) * dT/dH$, where H is depth, T is temperature, $k_{kond}(T)$ is the coefficient of thermal conductivity depending on temperature.

The one-dimensional stationary heat conduction equation is calculated numerically, or an analytical solution is found. The crustal thickness h_s is determined from the conditions $T(h_s) = T_L(h_s)$, T_L is the temperature of the phase transition of ice I into the liquid phase at pressure P at a depth h_s. Thus, the value of h_s is determined by the conditions on the Titan surface (temperature and heat flux F), the thermophysical properties of ice ($k_{kond}(T)$), and the phase diagram of ice Ih. Figure 2 shows the dependences of the thickness of the ice Ih shell on the surface heat flux F for different models of ocean composition.

For the chosen ocean model, the thickness of the Ih-crust and the structure of the water-ice shell [5-7] are uniquely determined by the magnitude of the heat flow. An increase on the flow F leads to a monotonic decrease in h_s. At low heat flow (below F~3.3 mW/m^2), the outer shell of Titan is represented only by ice (the inner ocean is not formed), while at higher heat flows, the satellite can form an inner ocean more than 400 km deep [6].

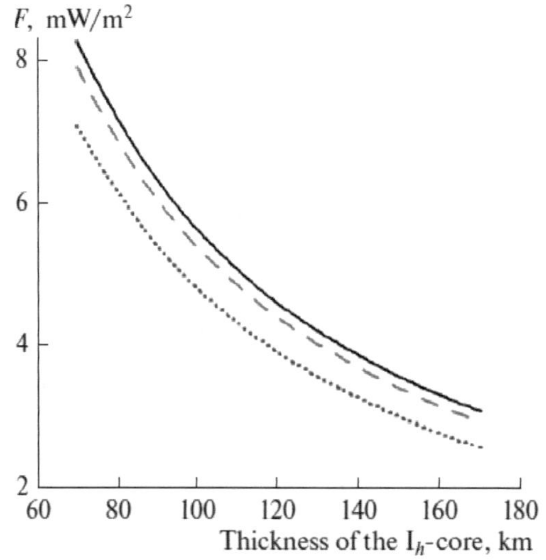

Figure 2. Thickness of the ice I*h* shell as a function of the surface heat flow *F* for different NH_3 concentrations in the ocean: 0% (solid line), 10% (dashed line), and 20% (dots).

5. Love numbers for a purely aquatic ocean models

For the obtained Titan models, the model Love numbers k_2 were determined. In fig. 3 shows the relationship between surface heat fluxes F, the corresponding thickness of the ice I_h-crust (h_s), and Love numbers k_2. The dependence of the model values of the k_2 Love numbers on the crustal thickness and, as a consequence, on the value of F is close to linear. The k_2 values increase with increasing F and decreasing h_s. Titan models without a water ocean are characterized by Love numbers almost an order of magnitude smaller than models with an ocean. The minimum Love numbers corresponding to the presence of an internal ocean in the satellite - $k_{2\min} \approx 0.44$.

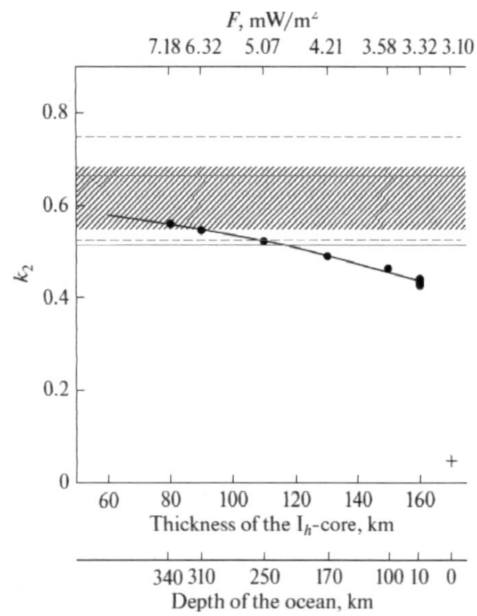

Figure 3. Dependence of Titan's Love number k_2 on the thickness of the ice I_h shell at different values of the surface heat flow of the satellite. The hatching shows the region of experimental values k_2 [3]. The solid and dashed lines are the allowable intervals of the Love numbers $k_2 \pm 1\sigma$ and $k_2 \pm 0.637$, respectively, according to the analysis of gravitational field data [2]. The cross in the lower part of the figures corresponds to values of the parameters k_2 in satellite models without the internal ocean.

6. Surface heat flow F from Titan

As shown above, to calculate the Love numbers in Titan, it is necessary to know the value of the surface heat flux F. Let us give estimates of the value of F.

Each calculated model of Titan can be represented as a quasi-stationary thermal model at a time instant (t), in which the surface heat flow of Titan F consists of the following components: the flow due to the radioactive decay of long-lived isotopes (F_R), the flow from dissipative processes (F_T), caused by tidal effects, and the flow from the energy of the phase transition (crystallization) released during the freezing of the inner ocean (F_C).

6.1 Radiogenic heat release

The magnitude of the radiogenic heat release of Titan F_R can be obtained from the heat release of chondrite matter for the time t from the formation of the solar system and the rock/water ratio in the satellite. Based on information on the radioactive properties of long-lived isotopes contained in chondrite matter (^{26}Al, ^{53}Mn, ^{60}Fe, ^{40}K, ^{232}Th, ^{235}U, ^{238}U), as well as data on the concentration of these elements in L/LL chondrites [14], the radiogenic heat release was estimated for this type of chondritic substance. Calculations showed that the radiogenic heating of the chondrite composition L/LL rock at the present time (t = 4.5 billion years) is $5.14 \cdot 10^{-12}$ W/kg. Knowing the mass of Titan and taking into account that the L/LL content of the chondrite component for the considered satellite model is about 50% [5], it is possible to estimate the contribution of radiogenic heat release to the surface heat flow F_R = 4.15 mW/m^2.

In alternative models of the internal structure of Titan, assuming complete differentiation of the satellite's material, the rock-iron core can be composed of hydrated low-density mineral material such as serpentine (antigorite), which is close in composition to the substance of CI chondrites [4, 15-16]. For such models, the values of the Love numbers are also determined mainly by the structure of the water-ice shell, the physical properties of ice I_h and the ocean [4, 13].

Therefore, for any composition of the rock-iron component (CI, L/LL, or mixed-type substance), the dependences of Love numbers and ice crust thickness on heat flow F in Titan's ice crust will be similar to those shown in Fig. 3, that is, the restrictions on the magnitude of surface heat fluxes in Titan are approximately the same for all types of chondrite material of the satellite. Our estimates of the radiogenic heating of the CI-rock, obtained from the data of [17], give a value of $3.46 \cdot 10^{-12}$ W/kg, which is 1.5 times less than the heat release in L/LL chondrite matter and, apparently, is the minimum heat generation for chondrites.

6.2 Tidal energy

The energy of dissipation (Q_T) or the energy of tidal heating depends on many parameters: the internal structure of Titan, the rheology of its interior, topography, characteristics of the satellite's dynamics, gravity fields in the Saturn system. Due to the complexity of the problem, Q_T estimates are currently available only for specific models. In particular, it was shown that the tidal dissipation in the cores of satellites of chondrite composition is negligible [18, 19]. The tidal energy released in the ocean (Q_{TO}) is estimated: $Q_{TO} \sim 1.19 \cdot 10^{10}$ W [20], $Q_{TO} \sim 1.1 \cdot 10^{10}$ W [21]. The Q_T value for an oceanless satellite (Q_{TS}) enriched in volatiles is determined to be $5.8 \cdot 10^{10}$ W and $6.43 \cdot 10^{10}$ W [21]. The maximum value of the satellite tidal energy can be approximately taken: $Q_T \sim Q_{TS} + Q_{TO} = 7.6 \cdot 10^{10}$ W, which corresponds to the contribution of the dissipation energy to the surface heat flow $F_T \sim 0.9$ mW/m^2.

6.3 Ice crystallization energy

To calculate the contribution of the crystallization energy, we use the approximate balance equations of surface fluxes F for two times t1, t2, which are solved iteratively. As a result of the solution, the models of the water-ice shells of Titan were obtained for the times t1 and t2. The energy release of ice crystallization has now been determined from the change in the volume of the ocean during the time t1-t2 and is estimated: $Q_C \sim 6.3 \cdot 10^{10}$ W.

6.4 Flow F for a purely aquatic ocean model

The surface heat flow F consists of the contribution of the dissipation energy ($F_T \sim 0.9$ mW/m^2), the contribution of the crystallization energy (0.76 mW/m^2), the contribution of the radioactive decay energy ($F_R = 4.15$ mW/m^2) and is F≈5.8 mW/m^2. The corresponding model of Titan has a crustal thickness of 100 km, an ocean depth of 280 km, $k_2 = 0.53$. Nevertheless, for complete agreement between the model and recently refined values of the Love numbers $k_2 = 0.616 \pm 0.067$ [3], it is necessary to have a surface flux $F \geq 6.3$ mW/m^2 or to adopt another model of the water-ice shell.

7. Volatiles in the water-ice shell of Titan

The composition of the water-ice phases in Titan is determined by the component composition of the ices delivered to the satellite by rocky-ice planetesimals and/or comets at different stages of its formation. Such ices mainly consist of H_2O ice (up to 75 wt%) and ices of other volatile compounds: CO (up to 20 wt%), CO_2 (up to 10 wt%), CH_3OH (up to 10 wt%), NH_3 (up to 3.5 wt%), CH_4 (up to 2 wt%), N_2 (up to 1 wt%) [22]. Titan's formation models [23] suggest the possibility of slow accretion of the satellite from small rock-ice particles. In this case, the temperature in the bowels of the growing satellite does not exceed the temperature of the phase transition of ice into liquid. However, at the later stages of satellite growth, the impact energy, due to the large mass of the growing satellite, leads to partial melting of ice, differentiation of ice and rock components, and, as a consequence, to the formation of a surface ocean of primary composition, in which volatile compounds are present in the form of dissolved gases or ions. As the ocean cools and crystallizes, the volatiles are redistributed between the ocean and the forming ice shell, into which they are incorporated as secondary chemically altered clathrate ice. Depending on the density ratio, the formed clathrates are distributed between the outer ice I_h-crust, the inner ocean, and the underlying high-pressure ice (HPI): light clathrates are part of the crust, the heavier ones remain in the ocean or are localized in the HPI region.

The composition of the water-ice phases in Titan is determined by the component composition of the ice delivered to the satellite. Substances with high solubility (ammonia, methanol) will remain dissolved, and their concentration in the water phase will increase as the ocean freezes and the ice thickness increases, but no more than up to 15 wt% [24]. In addition to volatile compounds transferred to the water ocean from the ice component of comets/planetesimals, the ocean composition could also be replenished with dissolved mineral salts due to the prolonged interaction of an aqueous solution with rocky (chondrite) matter deposited on the bottom of the primary ocean [25]. The equilibrium component composition of 5% of the ammonia ocean in contact with the CM chondrite rock at 5 kb has a density of 1209 g/L [25]. The density of CO_2 clathrate is higher than that of ice I_h, but lower or comparable to the density of seawater or ammonia-saline solution at ocean pressures in Titan. This allows us to assume the likely inclusion of CO_2 clathrates in the outer ice crust of the satellite. The elastic moduli (K_T and μ) of CO_2 clathrates in the pressure and temperature range close to the conditions of the outer I_h crust of Titan take values of 1.58-1.74 GPa [26], which is significantly less than μ of pure ice (4 GPa) [19] and should affect Love numbers.

8. Love numbers in models with volatiles in a water-ice shell

To estimate the Love numbers k_2 of Titan, we used the partial derivatives of the calculated Love numbers of the approximation dependence (1) with respect to the parameters of the water-ice shell. The input data are the results of calculating the Love number k_2 for a purely aquatic ocean model at F=5.8 mW/m^2, data on the density and rheological properties of the water-ice shell, Table 1, as well as phase diagrams of the ocean composition.

Calculations were carried out for various compositions of the aquatic ocean in Titan: pure or sea water, ammonia salt solution (5% NH_3 + salts), aqueous solutions of ammonia (15% NH_3) and methanol (10% CH_3OH). The calculation results in comparison with the k_2 data obtained from space observations are shown in Fig. 4. The measured values of k_2 are best satisfied by the model of Titan with an ocean of ammonia-salt composition and ice I_h-crust containing 5% CO_2 clathrates. The model that includes the ocean with sea water also gives a satisfactory agreement. Titan's models with a purely water ocean are within the error $k_2 \pm 1\sigma$. Calculations with water-ammonia and water-methanol oceans fall into the interval $k_2 \pm 1\sigma$ only in the case of sufficiently low (<5%) concentrations of dissolved NH_3 and CH_3OH.

Table 1. The parameters of the water-ice shell of Titan and the corresponding k_2 values of the satellite.

Model	Crust composition	Ocean composition	Density, kg/m^3		μ crust, GPa	k_2
			crust	ocean		
1	Ih +5%CO_2	Ammonia salt solution	943.57	1170	3.4	0.59
2	Ih	Sea water	936.11	1140	4.0	0.56
3	Ih	Pure water	936.11	1120	4.0	0.53
4	Ih	10% CH_3OH	936.11	1098	4.0	0.52
5	Ih	15% NH_3	936.11	1012	4.0	0.49

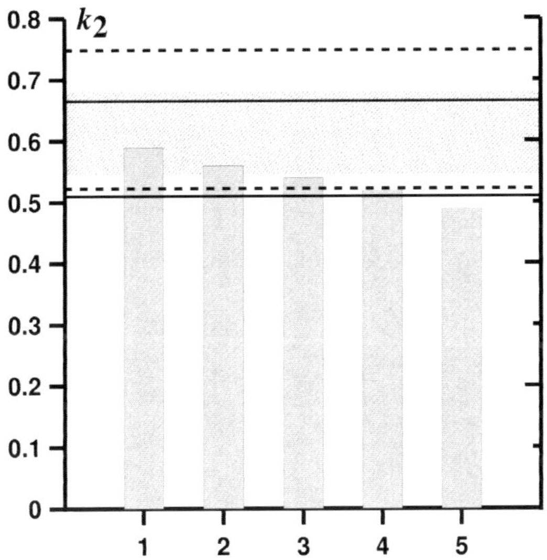

Figure 4. Love numbers for Titan with different compositions of the outer ice crust and the internal ocean of the satellite. 1 - ice crust composition: I_h + 5% CO_2, ocean composition: 5% NH_3 + dissolved salts; 2, 3, 4, 5 - Titan models with an icy I_h-crust and an internal ocean of the corresponding composition: seawater, pure water, methanol aqueous solution (10% CH_3OH), ammonia aqueous solution (15% NH_3). Hatching is the region of experimental k_2 values [3]. The solid and dotted lines are the $k_2 \pm 1\sigma$ values, respectively: 0.589 ± 0.075 and 0.637±0.112 [2].

9. Conclusion

The restrictions on the heat flux, the structure of the water-ice shell, and the composition of the Titan ocean, which are necessary to match the calculated (model) and experimental Love numbers k_2, are considered.

Estimates are obtained for model Love numbers and maximum surface heat fluxes for the L/LL chondrite composition of the rock-iron component of Titan, taking into account the radiogenic and tidal energy, as well as the energy of ice crystallization in the ocean due to satellite cooling.

The values of the tidal Love numbers k_2 of Titan were calculated for various models of the water-ice shell of the satellite, including the outer ice crust (ice I_h ± CO_2 clathrates) and the internal ocean of various compositions (pure or sea water, ammonia salt solution, aqueous solutions of ammonia and methanol). Titan models with an outer ice crust (ice I_h + 5% CO_2 clathrates) and an inner ocean representing a weakly concentrated saline solution based on an aqueous ammonia solution (5% NH_3) better than other tested models satisfy the experimental values of k_2.

Acknowledgments The work was performed as part of the state assignment of Vernadsky Institute of Geochemistry and Analytical Chemistry № 0137-2019-0004.

References

[1] Iess L et al. 2010 *Science* **327** (5971) 1367–69
[2] Iess L et al. 2012 *Science* **337** (6093) 457–59
[3] Durante D et. al 2019 *Icarus* **326** 123–32
[4] Baland R et al. 2014 *Icarus* **237** 29–41
[5] Dunaeva A N et al 2014 *Dokl. Earth Sci.* **454** 1 89–93
[6] Dunaeva A N et al. 2016 *Geochem. Int.* **54** 1 27–47
[7] Kronrod E V et al. 2020 *Solar System Research* **54** 405–419
[8] Wagner W. and Pruß A 2002 *Journal of physical and chemical reference data* **31** (2) 387–535
[9] Kuskov O L and Kronrod V A 2005 *Icarus.* **177** 550–569
[10] Kronrod V A and Kuskov O L 2003 *Geochemistry International* (9) 968–983
[11] Kuskov O L and Kronrod V A 2001 *Icarus* **151** 204–227
[12] Raevskiy S N et al 2015 *Izvestiya, Physics of the Solid Earth* (1) 139–147
[13] Beuthe M 2015 *Icarus* **258** 239–266.
[14] Wasson J T, Kalleymen G W 1988 Composition of chondrites. Philos. Trans. R. Soc. London A 325, 535–544
[15] Castillo-Rogez J C, Lunine J I 2010 *Geophys. Res. Lett.* **37** L20205.
[16] Sohl F et al 2014 *Journal of Geophysical Research: Planets* **119**(5) 1013–1036
[17] Braukmüller, N. et al. 2018. Geochimica et Cosmochimica Acta, 239, 17–48
[18] Sotin C and Tobie G 2004 *C. R. Phys.* **5**. 769–780
[19] Sohl F et al 2003 *J. Geophys. Res. Planets.* **108**(E12).
[20] Hay H C, Matsuyama I 2019 *Icarus* **319** 68–85
[21] Chen E et al 2014 *Icarus* **229** 11–30
[22] Marboeuf U et al 2014 *Astronomy & Astrophysics* **570** A35
[23] Barr A C et al 2010 *Icarus* **209**(2) 858–862
[24] Choukroun M and Grasset O 2010 *The Journal of chemical physics* **133**(14) 144502
[25] Leitner M A. and Lunine J I 2019 *Icarus* **333** 61–70
[26] Jia J et al 2017 *Scientific reports* **7**(1) 1–11

IOP Conf. Series: Materials Science and Engineering 1191 (2021) 012002 doi:10.1088/1757-899X/1191/1/012002

Analysis of Linear Stability and Bifurcations of Central Configurations in the Planar Restricted Circular Four-Body Problem

B S Bardin[1, 2] and E V Volkov[1, 2]

[1]Moscow Aviation Institute (National Research University), Volokolamskoe shosse 4, 125993 Moscow, Russia

[2]Mechanical Engineering Research Institute of the Russian Academy of Sciences, 4, M. Kharitonyevskiy Pereulok, 101990 Moscow, Russia

E-mail: bsbardin@yandex.ru, evvolkov94@mail.ru

Abstract. The planar restricted four-body problem is considered. That is, motion of an infinitesimal body P in the gravitational field of three attracting bodies is studied. It is supposed, that the above three body form a stable equilateral Lagrange triangle and all four bodies move in a plane. In this case there are eight central configurations formed by the bodies. An analysis of stability and bifurcation of the central configurations is performed. In particular, it is shown that the bifurcation is only possible in cases of degeneracy, when mass of an attracting body vanishes. It is also established that in non-degenerate cases five central configurations are unstable and three central configurations can be both stable and unstable. Domains of stability in linear approximation are constructed in the plane of parameters.

1. Introduction

The restricted four-body problem is a simplest generalization of the classical restricted three-body problem. It is of great interest for celestial mechanics and dynamics of satellites. In this problem, the motion of four bodies is investigated, one of which has an infinitely small mass, three other bodies have finite masses and move under the influence of their mutual attraction according to Newton's law of universal gravitation. The infinitesimal body moves under gravitational attraction of other three bodies and does not affect their motion. There exist a partial type of motion of the bodies, when they form the so-called central configuration. It means that the resultant attracting force acting on a body is a central force directed to the system mass center.

A numerous papers are devoted to the study of the central configuration in the restricted planar four-body problem. The existence and bifurcation of central configurations were studied in [1-5]. In particular, it has been established [1] that there can be eight, nine, or ten configurations. A linear analysis of the stability of central configurations has been performed in [6-9]. In [10-11], conclusions on the Lyapunov stability of central configurations were obtained in the case of two equal masses of attracting bodies.

In what follows, we denote by m_i masses of attracting bodies M_i (i = 1, 2, 3). We assume that the attracting bodies form an equilateral triangle (Lagrangian triangle) and move in circular orbits. The infinitesimal body P moves in the plane of the Lagrangian triangle. For any values of the masses of attracting bodies, we carry out a linear stability analysis and study possible scenarios of bifurcation for central configuration formed by small body *P* with attracting bodies M_i.

2. Formulation of the problem

Let us introduce rotating coordinate system *Oxyz*, which is rigidly connected with the Lagrangian triangle. The origin of this system is located in the middle of the segment M_2M_3. The axis *Ox* passes through this segment. The axis *Oy* is perpendicular to the *Ox* and passes through the attracting body M_1. The axis *Oz* complements the coordinate system to the right, orthogonal triple.

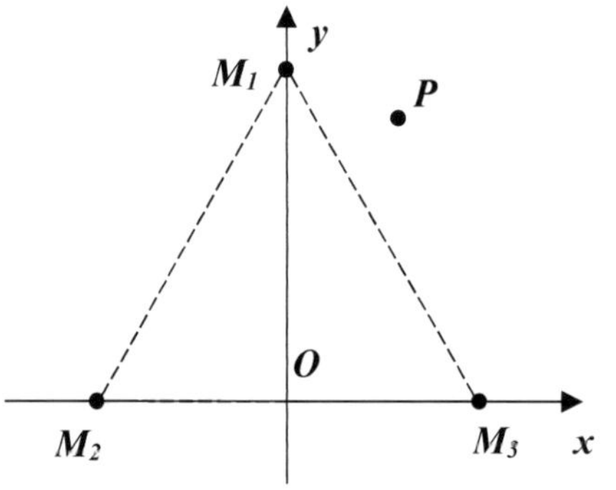

Figure 1. Coordinate system.

The equations of motion of the infinitesimal body *P* can be written in the form of Hamilton's equations

$$\frac{d\xi}{d\tau}=\frac{\partial H}{\partial p_\xi},\quad \frac{d\eta}{d\tau}=\frac{\partial H}{\partial p_\eta},\quad \frac{dp_\xi}{d\tau}=-\frac{\partial H}{\partial \xi},\quad \frac{dp_\eta}{d\tau}=-\frac{\partial H}{\partial \eta}. \tag{1}$$

The dimensionless coordinates ξ, η are introduced by means of formulae $x = r\xi$, $y = r\eta$, where *x* and *y* are coordinates of the infinitesimal body in system *Oxyz* and $r = const$ is the distance between the primaries. The dimensionless time τ is introduced by means of formula $\tau = \omega t$, where $\omega^2 = f(m_1 + m_2 + m_3)r^{-3}$ and *f* is the gravitational constant. Let us note that ω is the angular velocity of Lagrange triangular configuration.

The Hamiltonian function reads

$$H=\frac{1}{2}\left(p_\xi^2+p_\eta^2\right)+p_\xi\eta-p_\eta\xi-\frac{\sqrt{3}(1-\mu_2-\mu_3)}{2}p_\xi-\frac{\mu_2-\mu_3}{2}p_\eta-\frac{1-\mu_2-\mu_3}{\rho_1}-\frac{\mu_2}{\rho_2}-\frac{\mu_3}{\rho_3}, \tag{2}$$

where

$$\rho_1=\left(\xi^2+\left(\eta-\frac{\sqrt{3}}{2}\right)^2\right)^{1/2},\quad \rho_2=\left(\left(\xi+\frac{1}{2}\right)^2+\eta^2\right)^{1/2},\quad \rho_3=\left(\left(\xi-\frac{1}{2}\right)^2+\eta^2\right)^{1/2}. \tag{3}$$

$$\mu_2=\frac{m_2}{m_1+m_2+m_3},\quad \mu_3=\frac{m_3}{m_1+m_2+m_3}. \tag{4}$$

It is worth noting that the coordinates of the system mass center in dimensionless variables ξ, η read

$$\xi_c=-\frac{\mu_2-\mu_3}{2},\quad \eta_c=\frac{\sqrt{3}(1-\mu_2-\mu_3)}{2}. \tag{5}$$

In terms of the stability of the central configurations, we are only interested in values of the parameters μ_2 and μ_3, for which the Lagrangian triangular configuration of three bodies is stable. The necessary condition for the stability of such a configuration is the well-known Routh condition [12], which in terms of the parameters μ_2 and μ_3 reads

$$1 + 27(\mu_2^2 + \mu_2\mu_3 + \mu_3^2 - \mu_2 - \mu_3) > 0 \tag{6}$$

Without loss of any generality, we assume that the mass of the body M_1 is greater than the masses of the bodies M_2 and M_3. It follows from inequality (6) that the quantities μ_2 and μ_3 can take values from the interval

$$0 < \mu_i < \frac{1}{2} - \frac{\sqrt{69}}{18} \quad (i = 2, 3). \tag{7}$$

If the Routh condition is not satisfied, then the central configuration of bodies M_i (i = 1, 2, 3) is unstable. It yields the instability of the corresponding central configuration of four bodies. In what follows, we assume that the Routh condition is satisfied.

The equations of motion (1) have the following stationary solution describing the relative equilibria of infinitesimal body P in the rotating coordinate system $Oxyz$.

$$\xi = \xi_*, \quad \eta = \eta_*, \quad p_\xi = -\eta_* + \frac{\sqrt{3}}{2}(1 - \mu_2 - \mu_3), \quad p_\eta = \xi_* + \frac{1}{2}(\mu_2 - \mu_3), \tag{8}$$

where ξ_* and η_* satisfy the following equations

$$\begin{aligned} &\xi + \frac{\mu_2}{2} - \frac{\mu_3}{2} - \frac{(1 - \mu_2 - \mu_3)}{\rho_1^3}\xi - \frac{\mu_2}{\rho_2^3}\left(\xi + \frac{1}{2}\right) - \frac{\mu_3}{\rho_3^3}\left(\xi - \frac{1}{2}\right) = 0, \\ &\eta - \frac{\sqrt{3}(1 - \mu_2 - \mu_3)}{2} - \frac{(1 - \mu_2 - \mu_3)}{\rho_1^3}\left(\eta - \frac{\sqrt{3}}{2}\right) - \frac{\mu_2}{\rho_2^3}\eta - \frac{\mu_3}{\rho_3^3}\eta = 0. \end{aligned} \tag{9}$$

The equilibrium of the body P in the rotating coordinate system $Oxyz$ corresponds to its absolute motion in a circular orbit. In this motion the body P forms a constant central configuration with the bodies M_i.

In the next sections for any parameters values we investigate linear stability of the corresponding central configuration and perform analysis of their bifurcations.

3. Bifurcation of relative equilibria

Let us first consider the limiting cases $m_2 = 0$ and $m_3 = 0$, when our problem degenerates into the restricted three-body problem. In this cases, the central configurations degenerate into Euler and Lagrangian libration points. The central configurations for the limiting cases are shown in figures 2 and 3. The positions of the body P is indicated by blue points. To designate the relative equilibrium P in limiting cases, we used the standard notation for libration points $L_j^{(k)}$, where the superscript k corresponds to the case $\mu_k = 0$.

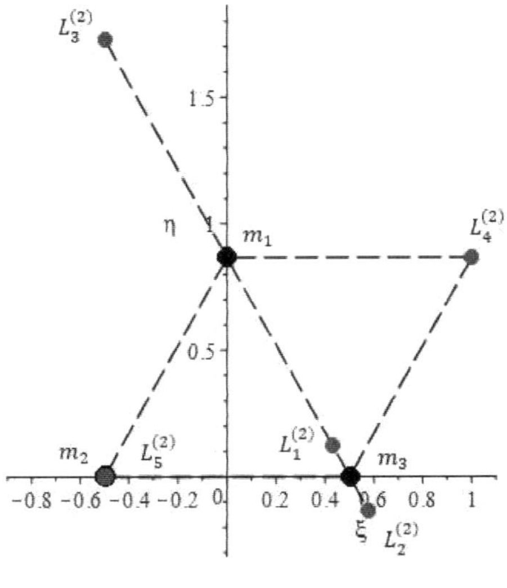

Figure 2. Central configurations in the limiting case $\mu_2 = 0$ ($m_2 = 0$).

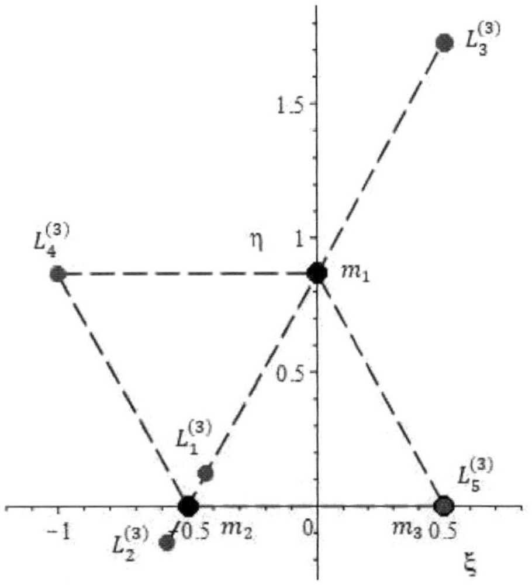

Figure 3. Central configurations in the limiting case $\mu_3 = 0$ ($m_3 = 0$).

For parameter values satisfying the Routh condition, the regions of relative equilibria for the infinitesimal body P were numerically constructed (see figure 4). It turned out that for parameter values $\mu_2 \neq 0$ and $\mu_3 \neq 0$ there are exactly eight relative equilibria, which are located in narrow regions. These regions are located either along a circle of unit radius, whose center is in the body M_1 or along the straight lines passing through M_1, M_3 and M_1, M_2 respectively (see figure 4). The libration points $L_j^{(k)}$ are the limit points of the indicated regions, so that the relative equilibria pass into one of the libration points at $\mu_2 = 0$ or $\mu_3 = 0$.

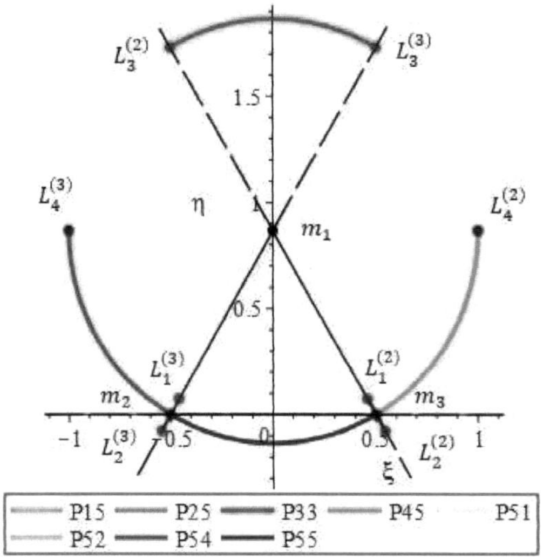

Figure 4. Regions of relative equilibria.

By P_{ij} we denote the relative equilibrium that degenerates into a libration point $L_i^{(2)}$ as $\mu_2 \to 0$ and degenerates into a libration point $L_j^{(3)}$ as $\mu_3 \to 0$. The relative equilibria P_{ij} and corresponding libration points are shown in table 1.

Table 1. Relative equilibria and their limiting positions.

$\mu_2 = 0$ $(m_2 = 0)$	$\mu_2 \neq 0$, $\mu_3 \neq 0$	$\mu_3 = 0$ $(m_3 = 0)$
$L_1^{(2)}$	$\boldsymbol{P_{15}}$	$L_5^{(3)}$
$L_2^{(2)}$	$\boldsymbol{P_{25}}$	$L_5^{(3)}$
$L_3^{(2)}$	$\boldsymbol{P_{33}}$	$L_3^{(3)}$
$L_4^{(2)}$	$\boldsymbol{P_{45}}$	$L_5^{(3)}$
$L_5^{(2)}$	$\boldsymbol{P_{51}}$	$L_1^{(3)}$
$L_5^{(2)}$	$\boldsymbol{P_{52}}$	$L_2^{(3)}$
$L_5^{(2)}$	$\boldsymbol{P_{54}}$	$L_4^{(3)}$
$L_5^{(2)}$	$\boldsymbol{P_{55}}$	$L_5^{(3)}$

The regions of relative equilibria P_{ij} have no common points except for the limiting positions $L_5^{(2)}$ and $L_5^{(3)}$ corresponding to the cases $\mu_2 = 0$ and $\mu_3 = 0$. At $\mu_k = 0$ $(k = 2, 3)$ bifurcations of relative equilibria can occur. In fact, the following scenarios of bifurcations take place. The relative equilibria P_{51}, P_{52}, P_{54} and P_{55} approach $L_5^{(2)}$ as $\mu_2 \to 0$ and merge into the point $L_5^{(2)}$ at $\mu_2 = 0$. Similarly, the relative equilibria P_{15}, P_{25}, P_{45} and P_{55} approach $L_5^{(3)}$ as $\mu_3 \to 0$ and merge into the point $L_5^{(3)}$ at $\mu_3 = 0$. Regions of relative equilibria near $L_5^{(2)}$ and $L_5^{(3)}$ are shown in figures 5, 6.

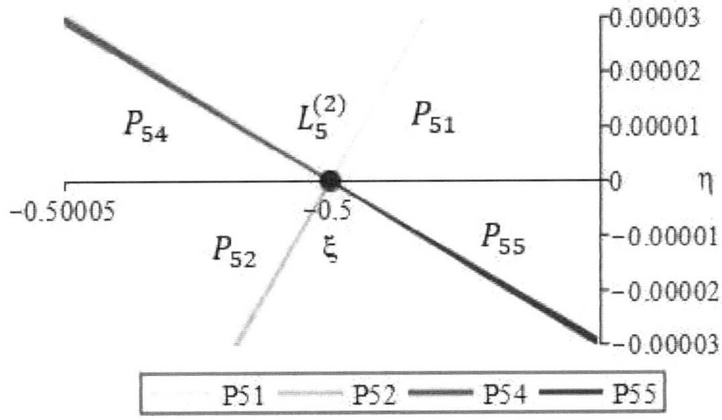

Figure 5. The fragments of regions for existence of relative equilibria near the point $L_5^{(2)}$.

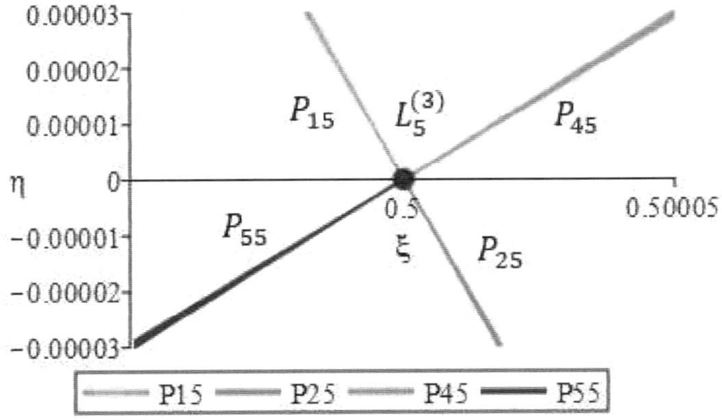

Figure 6. The fragments of regions for existence of relative equilibria near the point $L_5^{(3)}$.

From the above numerical analysis, it is clear that bifurcations of relative equilibria are possible only in the limiting cases mentioned.

Note also that the regions of existence of relative equilibria were constructed using the method of numerical continuation in the parameters. In the vicinity of the points $L_5^{(2)}$ and $L_5^{(3)}$, the numerical analysis becomes more complicated and requires higher accuracy of calculations. Let us note that in the vicinity of the points $L_5^{(2)}$ and $L_5^{(3)}$ the study of the relative equilibria can be carried out analytically by using the small parameter method, which in this case is more efficient than the numerical analysis.

4. Stability of central configuration

In this section, we present the results of studying the stability of central configurations in linear approximation. If the Routh condition is satisfied, then the problem of stability of the central configuration is reduced to studying the stability of relative equilibria of the body P in the coordinate system $O\xi\eta$ rotating together with the attracting centers.

In order to study the stability of equilibria P_{ij} we introduce local canonical variables q_1, q_2, p_1 and p_2 in a neighborhood of the above equilibria

$$\xi = \xi_* + q_1, \quad \eta = \eta_* + q_2,$$
$$p_\xi = -\eta_* + \frac{\sqrt{3}}{2}(1 - \mu_2 - \mu_3) + p_1, \quad p_\eta = \xi_* + \frac{1}{2}(\mu_2 - \mu_3) + p_2, \tag{10}$$

and expand the Hamiltonian (2) in a series with respect to powers of the new canonical variables

$$H = H_2 + H_3 + H_4 + \cdots. \tag{11}$$

The quadratic part H_2 of the Hamiltonian reads

$$H_2 = \frac{1}{2}(p_1^2 + p_2^2) + \frac{1}{2}(aq_1^2 + bq_2^2) + cq_1q_2 + p_1q_2 - p_2q_1, \tag{12}$$

where the coefficients a, b, c have the following explicit form

$$a = -\frac{(1-\mu_2-\mu_3)}{\rho_1^5}(3\xi_*^2 - \rho_1^2) + \frac{\mu_2}{\rho_2^5}\left(\rho_2^2 - 3\left(\xi_* + \frac{1}{2}\right)^2\right) + \frac{\mu_3}{\rho_3^5}\left(\rho_3^2 - 3\left(\xi_* - \frac{1}{2}\right)^2\right),$$
$$b = -\frac{(1-\mu_2-\mu_3)}{\rho_1^5}\left(3\left(\eta_* - \frac{\sqrt{3}}{2}\right)^2 - \rho_1^2\right) + \frac{\mu_2}{\rho_2^5}(\rho_2^2 - 3\eta_*^2) + \frac{\mu_3}{\rho_3^5}(\rho_3^2 - 3\eta_*^2), \tag{13}$$
$$c = -\frac{3(1-\mu_2-\mu_3)}{\rho_1^5}\xi_*\left(\eta_* - \frac{\sqrt{3}}{2}\right) - \frac{3\mu_2}{\rho_2^5}\left(\xi_* + \frac{1}{2}\right)\eta_* - \frac{3\mu_3}{\rho_3^5}\left(\xi_* - \frac{1}{2}\right)\eta_*.$$

The question of the stability for this linear system with the Hamiltonian H_2 can be solved by analysing the roots of its characteristic equation

$$\lambda^4 + (a + b + 2)\lambda^2 + ab - c^2 - a - b + 1 = 0. \tag{14}$$

In accordance with Lyapunov's theorem [13], the linear system is stable if all roots of the characteristic equation are simple and purely imaginary. This condition satisfied if the coefficients of equation (14) satisfy the following inequalities

$$a + b + 2 > 0,$$
$$ab - c^2 - a - b + 1 > 0, \tag{15}$$
$$a^2 + b^2 + 4c^2 - 2ab + 8(a + b) > 0.$$

Conditions (15) were verified numerically for all possible values of the parameters μ_2, μ_3 and the following conclusions on stability were obtained. The relative equilibrium P_{33} is unstable for all values of the parameters μ_2 and μ_3. The relative equilibria P_{15} and P_{25} are unstable when $\mu_3 \neq 0$. In the limiting case $\mu_3 = 0$ the equilibria P_{15} and P_{25} degenerate into the libration point $L_5^{(3)}$, which is stable in linear approximation for all values of the parameter μ_2. Similarly, the relative equilibria P_{51}

and P_{52} are unstable when $\mu_2 \neq 0$. If $\mu_2 = 0$, then P_{51} and P_{52} degenerate into the libration point $L_5^{(2)}$, which is stable in linear approximation for all values of the parameter μ_3.

The most interesting situation takes place for the relative equilibria P_{45}, P_{54} and P_{55}. These relative equilibria, depending on the values of the parameters, can be both stable and unstable. For these relative equilibria stability diagrams were constructed in the plane of the parameters (see figures 7, 8, 9). The instability domains are indicated by red color and domains of stability the linear approximation are indicated by blue color.

Note that in the limiting case $\mu_2 = 0$ the relative equilibria P_{45}, P_{54} and P_{55} degenerate into the libration point $L_5^{(2)}$, which is stable in linear approximation for all values of the parameter μ_3. Similar, in the limiting case $\mu_3 = 0$ they degenerate into the point of libration $L_5^{(3)}$, which is stable in linear approximation for all values of the parameter μ_2.

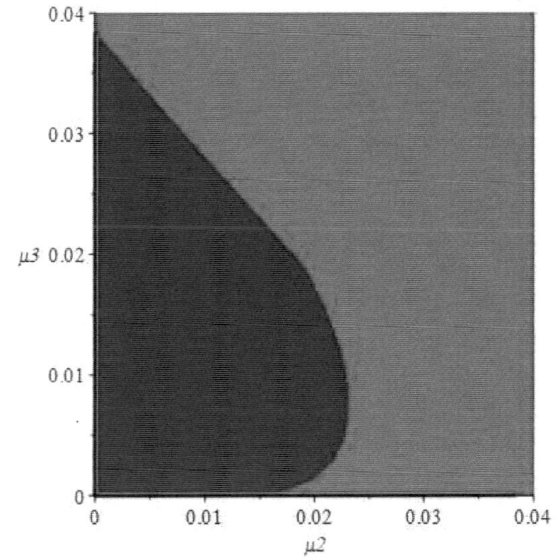

Figure 7. Domains of stability of relative equilibria P_{45}.

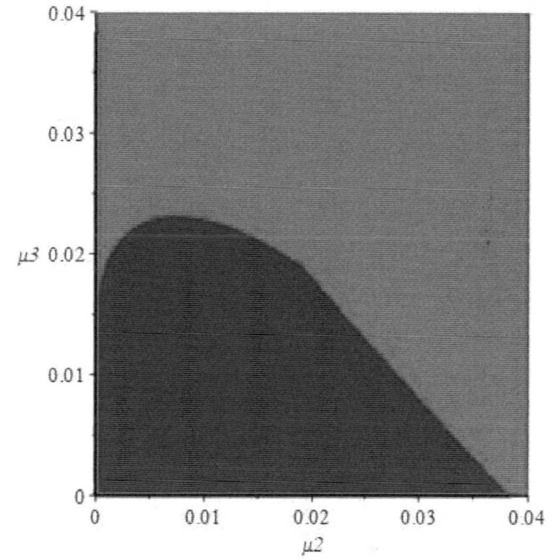

Figure 8. Domains of stability of relative equilibria P_{54}.

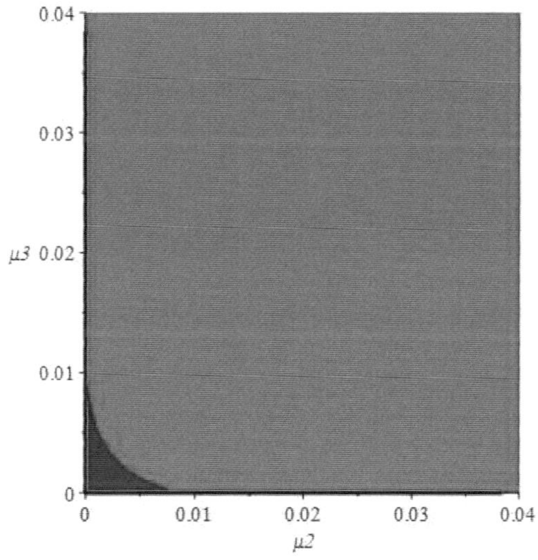

Figure 9. Domains of stability of relative equilibria P_{55}.

Conclusions

Let us briefly formulate the results of the study. It has been numerically established that if Routh conditions are satisfied, then there exist exactly eight central configurations, which correspond to the relative equilibria of the infinitesimal body P in the coordinate system rotating together with the attracting centers M_1, M_2 and M_3. In the limiting cases, when an attracting center has zero mass, the relative equilibria degenerate into triangular (Lagrangian) or rectilinear (Euler) libration points. In this case, bifurcation is possible. It occurs according to the following scenarios. Four relative equilibria P_{51}, P_{52}, P_{54} and P_{55} pass into the libration point $L_5^{(2)}$ at $\mu_2 = 0$, and four other relative equilibria P_{15}, P_{25}, P_{45} and P_{55} pass into the libration point $L_5^{(3)}$ at $\mu_3 = 0$. There are no other bifurcations.

The regions of existence of possible relative equilibria of the infinitesimal body are numerically constructed for all values of the parameters that satisfy the Routh conditions. By using the numerical analysis, it was established that the position of relative equilibrium P_{33} is unstable for all values of the parameters. The relative equilibria P_{15} and P_{25} are linearly stable only in the limiting case $\mu_3 = 0$, when they pass into the libration point $L_5^{(3)}$, and P_{51} and P_{52} are linearly stable only in the limiting case $\mu_2 = 0$, when they pass into the libration point $L_5^{(2)}$. The relative equilibria P_{45}, P_{54} and P_{55} can be both stable and unstable depending on value of parameters.

Acknowledgments

This work was supported by the grant of the Russian Scientific Foundation (project No. 19-11-00116) at the Moscow Aviation Institute (National Research University).

References

[1] Pedersen P 1944 *Dan. Mat.-Fys. Medd* **21** 1-80
[2] Pedersen P 1952 *Dan. Mat.-Fys. Medd* **26** 1-37
[3] Budzko D A and Prokopenya A N 2010 *Programming and Computer Software* **36** 68-74
[4] Leandro E S G 2006 *J. Differential Equations* **226** 323–351
[5] Howell K C and Spencer D B 1986 *Acta Astronomica* **13** 473-479
[6] Brumberg V A 1957 *Soviet Astronomy* **1** 57–79
[7] Budzko D A 2009 *Computer algebra systems in teaching and research* **1** 28-36
[8] Grebenikov E A, Gadomski L and Prokopenya A N 2007 *Nonlinear Oscillations* **10** 62–77
[9] Budzko D A 2011 *Bull. of the National Academy of Sciences of Belarus* **4** 55-59
[10] Bardin B S and Volkov E V 2020 Stability Study of a Relative Equilibrium in the Planar Circular Restricted Four-Body Problem *IOP Conf. Ser.: Materials Science and Engineering* **927** 012012
[11] Bardin B S and Esipov P A 2018 Investigation of Lyapunov stability of a central configuration in the restricted four-body problem *AIP Conf. Proc.* **1959** 040004
[12] Routh E J 1875 *Proc. London Math. Soc.* **6** 86–97
[13] Lyapunov A M 1992 *The General Problem of the Stability of Motion* (London: Taylor & Francis)

IOP Conf. Series: Materials Science and Engineering 1191 (2021) 012003 doi:10.1088/1757-899X/1191/1/012003

Modeling of dynamic interaction of a fabric airbag with an anthropometric dummy

V V Selivanov, D P Levin, A V Petukov, M E Arcivenko and A V Gonsales Astua

Bauman Moscow State Technical University, 5/1, 2-nd Baumanskaya, Moscow, 105005, Russia

E-mail: dlevine@yandex.ru

Abstract. In this paper, a physical and mathematical model of interaction between a fabric expanding automobile airbag and biological objects is formulated. Bioobjects are represented using finite element models of anthropometric dummies of the Hybrid III 50th Male type. As a model of the airbag material, a model of multi-layer fabric orthotropic material was used. The method of corpuscular particles was used for numerical simulation of the airbag functioning in the LS-DYNA multiphysics simulation software package. The results of calculating the dynamic interaction of an expanding airbag with three standing dummies are presented.

1. Introduction

The purpose of this work is to develop a method for numerical modeling of the dynamic interaction of a tissue expanding air-filled object with mechanical models of biological objects. Such objects (airbags) are widely used as the main elements of the passive safety system of modern cars – SRS (Supplementary Restraint System), which provides protection for the driver and passengers in a head-on collision with a static or moving object.

The airbag module of commercial driver and passenger safety systems includes the airbag itself (a thin, multi-layered, nylon shell with a thickness of no more than 0.4 mm), and its associated actuation mechanisms: a pyropatron and a gas generator. The response speed, i.e. the time to fully open the airbag, is no more than 20 ms. The pressure in the airbag is maintained for 10 seconds, after which the gas is drained through a perforation in the fabric of the airbag.

The specifics of the design scheme considered in this paper are determined by the possible use of an airbag as a trauma-safe striker in non-lethal weapon systems [1-3].

2. Calculation scheme and models

Consider the following calculation scheme: in a room where there are several people who are subject to temporary neutralization of their mobility, a rolled-up airbag with a surface area of 1 m^2, having an initial speed of about 50 m/s, is triggered (Fig. 1).

The problem was solved numerically by the Lagrangian method implemented in the LS-DYNA program. It is based on a system of equations of the laws of conservation of mass, momentum and energy:

$$\rho V = \rho_0 \qquad \sigma_{ij,j} + \rho f_i = \rho \ddot{x}_i \qquad (1)$$
$$\dot{E} = V s_{ij}\dot{\varepsilon}_{ij} - (p+q)\dot{V} \qquad s_{ij} = \sigma_{ij} + (p+q)\delta_{ij}$$

where $\sigma_{ij,j}$ – components of the Cauchy stress tensor divergence; s_{ij} – components of the stress tensor deviator, ρ and ρ_0 – current and initial density of the material, respectively; f_i – components of the vector of volumetric forces acting the body; $\ddot{x}_i$ – acceleration components of the element; $\dot{E}$ – total time derivative of specific internal energy; V and $\dot{V}$ – the relative specific volume and its total time derivative, respectively; $\dot{\varepsilon}_{ij}$ – components of the strain rate tensor; p – current pressure; q – additive to the pressure due to artificial viscosity; δ_{ij} – Kronecker Delta function. This system of equations is supplemented by strength models and equations of state of the materials used in the calculations [4].

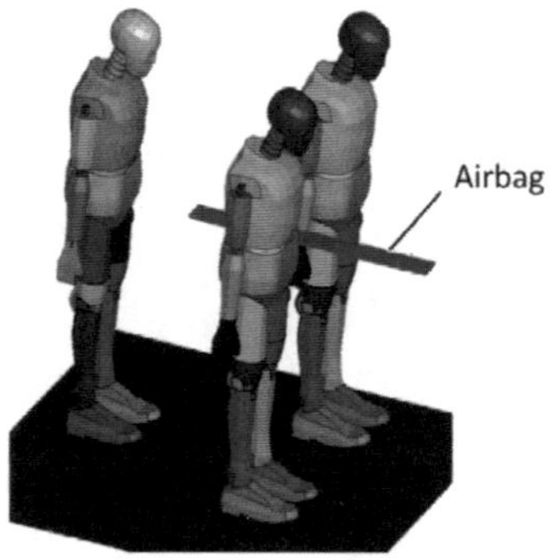

Figure 1. Calculation scheme

As a model of the airbag material, a model of multi-layer fabric orthotropic material was used [5]. Each calculated cell of the airbag had its own orthogonal coordinate system (Fig. 2). In this case, the c axis always coincided with the normal to the cell, and the a and b axes lay in the plane of the cell.

The relationship between stress and strain was defined as

$$C = T^T C_L T \tag{2}$$

where T – transformation matrix, and C_L – defining matrix defined by material constants for orthogonal axes of the material a, b and c. The inverse matrix for the C_L in the orthotropic case was given as follows:

$$C_L^{-1} = \begin{vmatrix} \frac{1}{E_a} & -\frac{v_{ba}}{E_b} & -\frac{v_{ca}}{E_c} & 0 & 0 & 0 \\ -\frac{v_{ba}}{E_a} & \frac{1}{E_b} & -\frac{v_{cb}}{E_c} & 0 & 0 & 0 \\ -\frac{v_{ca}}{E_a} & -\frac{v_{cb}}{E_b} & \frac{1}{E_c} & 0 & 0 & 0 \\ 0 & 0 & 0 & \frac{1}{G_{ab}} & 0 & 0 \\ 0 & 0 & 0 & 0 & G_{bc} & 0 \\ 0 & 0 & 0 & 0 & 0 & G_{ca} \end{vmatrix}, \tag{3}$$

$$\frac{v_{ab}}{E_a} = \frac{v_{ba}}{E_b}; \frac{v_{ca}}{E_c} = \frac{v_{ac}}{E_a}; \frac{v_{cb}}{E_c} = \frac{v_{bc}}{E_b}$$

where E_a, E_b, E_c – Young's modules in the direction of the corresponding axes of the material; v_{ba}, v_{ca}, v_{cb} – Poisson coefficients in the coordinate planes of the material; G_{ab}, G_{bc}, G_{ca} – shear modulus in the coordinate planes of the material. Table 1 shows the constants for this model used in the calculation.

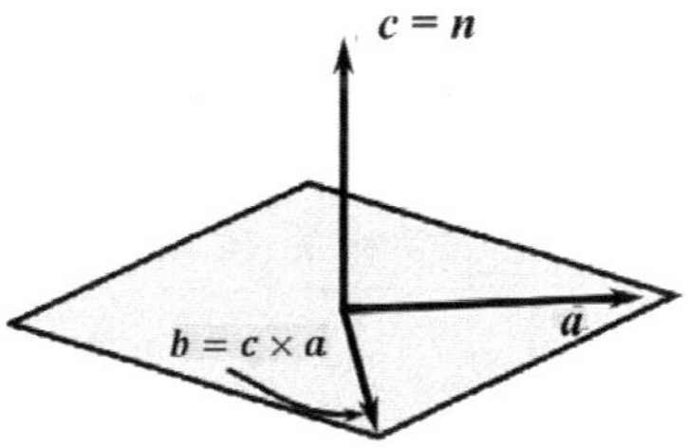

Figure 2. Orthogonal coordinate system in model cell

Table 1. The coefficients for the model of the strength of the airbag

Density, kg/m^3	E_a, GPa	E_b, GPa	ν_{ba}	ν_{ca}	G_{ab}, GPa
2000	2	2	0,35	0,35	2

Modeling of biological objects in the room was performed using finite element models of anthropometric dummies of the Hybrid III 50th Male type [6] (Fig. 3). The weight of one dummy was 95 kg.

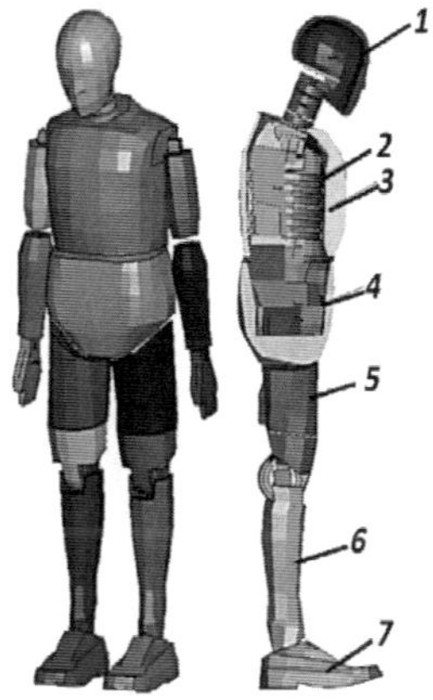

Figure 3. Dummy Hybrid III 50th Male: 1-skull; 2-ribs; 3- thorax; 4-pelvis; 5-thigh; 6-shin; 7-foot

The hip, lower leg, and foot were described by a model of a bilinear elastic-plastic body. A viscoelastic model [5] was chosen as the material model for the skull and ribs. The relaxation of the shear modulus obeyed the equation

$$G(t) = G_\infty + (G_o - G_\infty)e^{-\beta t} \tag{4}$$

where G_∞ – shear modulus under prolonged loading; G_0 – short-term shear modulus; β – attenuation constant.

To describe the materials of the chest and pelvis, a porous body model was used, in which the relationship between stresses and deformations was described by a nonlinear relationship during compression (Fig. 4), and a linear relationship between stresses and deformations was assumed during tension [5].

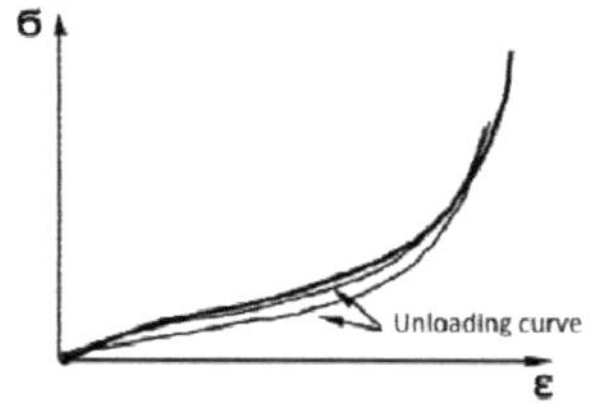

Figure 4. Relationship between stresses and deformations

3. Models verification

Specific constants of materials were selected in such a way as to most accurately match the data of calibration tests: head impact, thorax impact [7-9]. The interaction between different parts of the dummies was described using a penalty contact algorithm. When using this algorithm, the repulsive force of touching cells of materials is directly proportional to the depth of their interpenetration (Fig. 5). The proportionality coefficient was defined as

$$K = \frac{\rho c A}{\Delta t} \tag{5}$$

where ρ – material density; c – sound speed; A –area of a cell; Δt – time integration step.

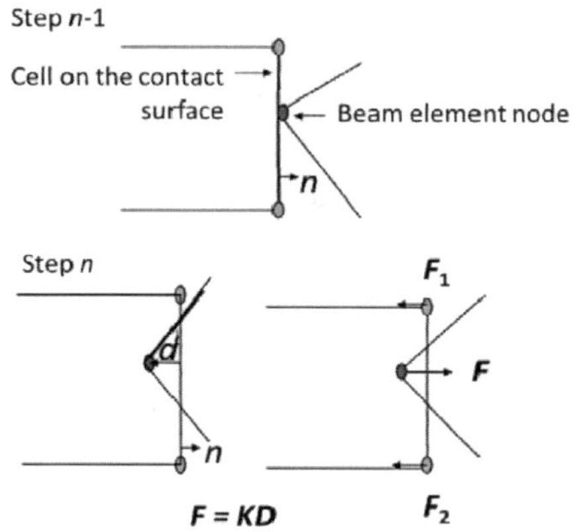

Figure 5. Scheme of contact interaction

Figure 6 shows the calculation scheme of the test, where steel cylinder impact the dummy chest with a velocity of 6.71±0.12 m/s. The direction of impact coincides with the axis of rotation of the cylinder and the plane of symmetry of the dummy, and the impact point is located at a distance of 12.7 mm from the middle surface of the third rib of the dummy.

Figure 7 and table 2 show the main results of the calculation. According them, computational model describes the experimental data well (all numerical results are within the experimental error).

Table 2. Comparison of the results of calculating the impact to the chest with the experiment

Parameter	Experimental results	Numerical results
Cylinder velocity, m/s	6,71±0.12	6,68
Thorax deflection, mm	68,05±4.55	68,1
Peak resistance force, N	5527±367	5196,2
Internal hysteresis	77±8%	75,7%

Impact of the head on the steel plate was also calculated for dummy testing (Fig. 8). In the initial moment of time the head of the dummy is suspended on holding devices at a distance of 376 mm from a rigid surface. Then, under the influence of gravity, it falls and hits. The impact velocity reaches 6.71 m/s.

Table 3 shows the main results of the calculation. It also shows a good agreement between the calculated and experimental data, the calculation error is no more than 8%.

Figure 6. Scheme of the thorax impact test

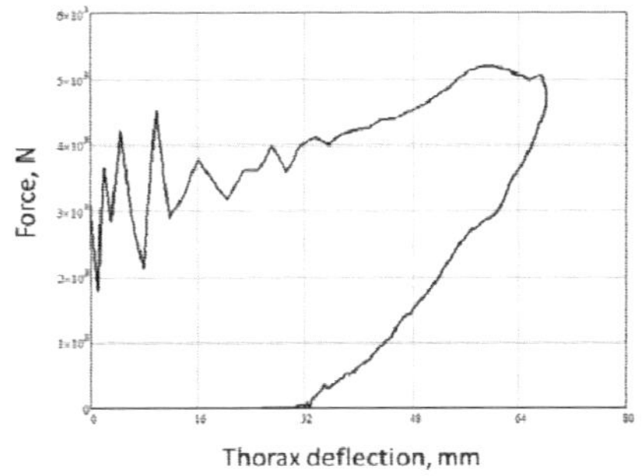

Figure 7. The dependence of the contact force from the deflection of the thorax

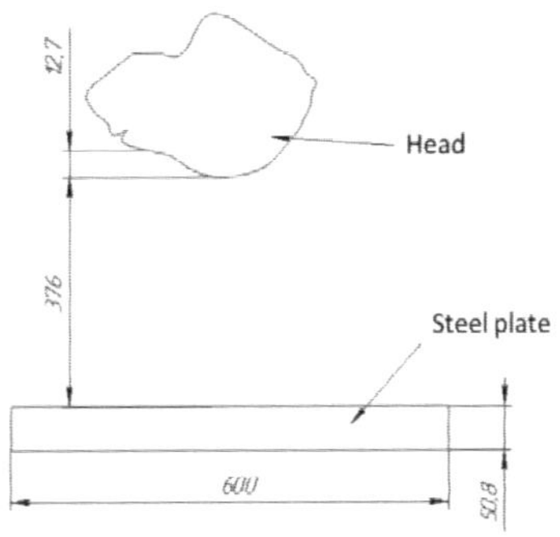

Figure 8. Scheme of the head impact test

Table 3. Comparison of the results of calculating the impact to the head with the experiment

	Experimental results	Numerical results	Error, %
Peak acceleration, m/s^2	10463,7	11223,3	7,26
Head mass, kg	4,54±0,05	4,54	-
Moment of inertia, mm·kN·ms^2	23,388±11,239	22,405	4,2

4. Method of corpuscular particles

Until now, one of the main methods for modeling airbag opening has been the control volume method [5]. In accordance with the algorithm of this method, the airbag was considered as a control surface modeled by a set of shell finite elements. The area of the control surface was associated with the internal volume using Green's formula:

$$\iiint \varphi \frac{\partial \psi}{\partial x} dxdydz = -\iiint \psi \frac{\partial \varphi}{\partial x} dxdydz + \oint \psi \varphi n_x d\Gamma \qquad (6)$$

IOP Conf. Series: Materials Science and Engineering 1191 (2021) 012003 doi:10.1088/1757-899X/1191/1/012003

where the first two integrals are integrals over a closed volume, and the last integral is an integral over the surface covering that volume; n_x – Is the guiding cosine between the normal to the surface and the x direction.

The volume was calculated as

$$V = \iiint dxdydz\,. \qquad (7)$$

By defining arbitrary functions φ and ψ as $\varphi = 1$ and $\psi = x_x$.

Finally we get:

$$V = \iiint dxdydz = \oint xn_x d\Gamma\,. \qquad (8)$$

The surface integral can be approximated by the sum over all shell elements that make up the airbag:

$$\oint xn_x d\Gamma \approx \sum_{i=1}^{N} \overline{x_i} n_{ix} A_i \qquad (9)$$

where $\overline{x_i}$ is the average coordinate of the calculated cell; n_{ix} – the guiding cosine between the normal to the cell and the x direction; A_i – the area of the cell.

The pressure corresponding to the current volume was determined from the equation of state of the ideal gas and applied simultaneously to all the cells of the airbag (Fig. 9).

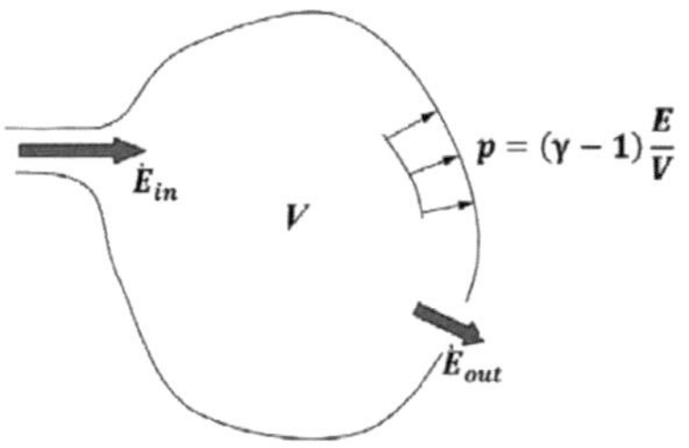

Figure 9. Pressure application diagram for the control volume method

The main advantages of this method are its speed, stability and good convergence, and the disadvantage is incorrect modeling of situations in which the interaction of the airbag and objects occurs before it is fully deployed.

In this work, we used the method of corpuscular particles [7], which allows us to neutralize this disadvantage. In contrast to the method described above, the airbag inflates under the action of a stream of spherical particles, which are essentially ensembles of many real gas molecules (Fig. 10). This ensures that the specific translational kinetic energy of molecules and particles is equal. The interactions between the particles themselves and the fabric were perfectly elastic collisions (Fig. 11).

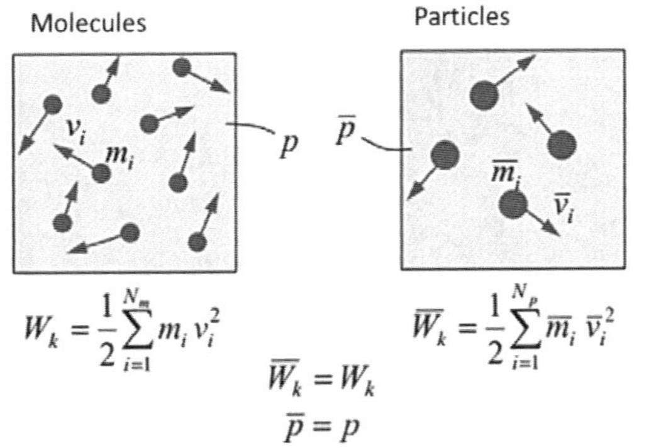

Figure 10. Method of corpuscular particles [7]

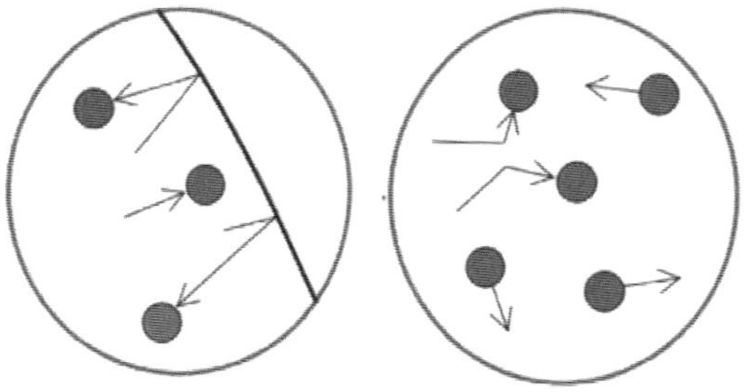

Figure 11. Diagrams of interactions of gas particles with the airbag fabric (left) and with other particles (right) [7]

5. Results and discussion

Figure 12 illustrates the process of inflating the airbag. At the initial moment of time, the gas generator is activated, and the gas begins to flow into the upper layer, which inflates and crushes the remaining layers. Then there is a reversal of the inflated layer and the gas begins to flow into the folded part of the airbag.

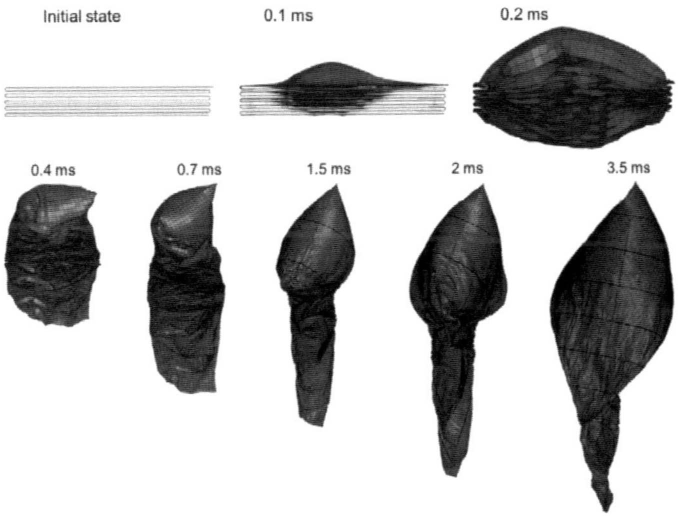

Figure 12. Inflating of the airbag

Figure 13 shows the interaction between the airbag and the dummies. Under the influence of the initial velocity, the airbag, which has not yet fully opened, hits the dummy on the far left in the thorax and pelvis area. As a result of the exchange of impulses, the speed is damped and the cushion rebounds, as well as the speed is set by the dummy. After that, the airbag falls to the floor, bounces off it and hits the leg of the second dummy.

Figure 14 shows graphs of the speed set of the thorax and pelvis of the dummy, on which the initial impact is made.

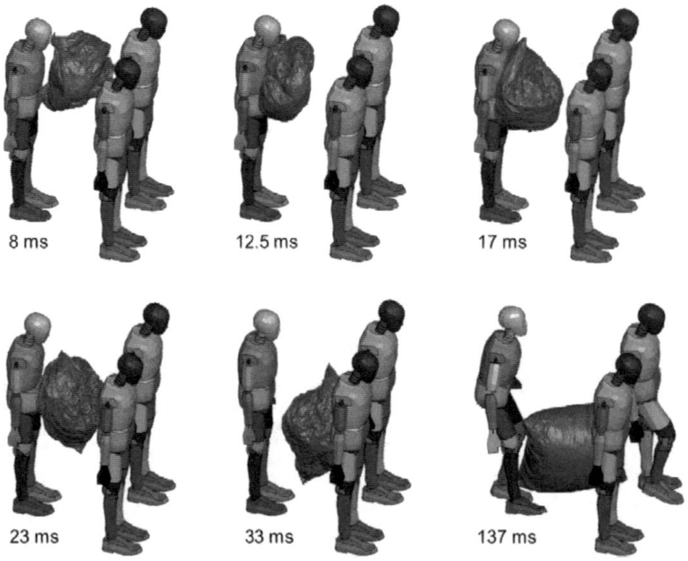

Figure 13. Interaction between airbag and dummies

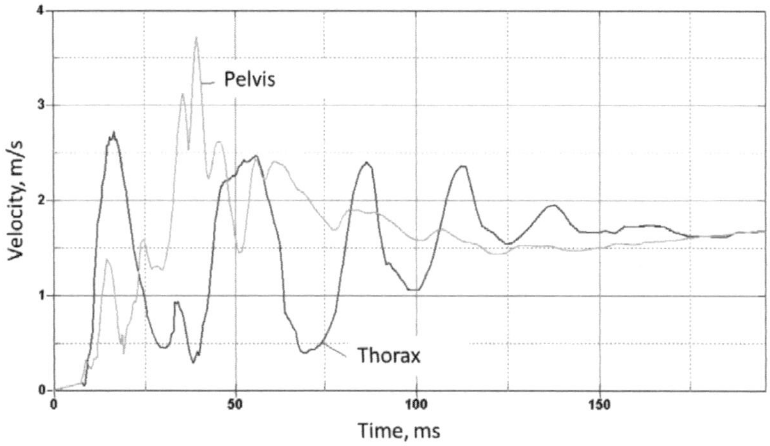

Figure 14. Dependence of the speed of the thorax and pelvis of the dummy on time

As a result of the calculations, it was shown that this approach allows modeling the deployment of a collapsed airbag with sufficient accuracy for engineering applications, as well as determining the loads (pressure, momentum, acceleration distributions) that impact various parts of anthropometric dummies.

References

[1] Levin D P, Selivanov V V 2018 *Defense Eng. Problems. Ser. 16* **11-12** 119-126 (in Russian)
[2] Levin D P, Lyushnin S A 2013 *Eng. bulletin* **10** ainjournal.ru/doc/631502.html (*in Russian*)
[3] Levin D P, Selivanov V V, Lyushnin S A 2012 *Defense Eng. Problems. Ser. 16* **7-8** 63-76 (*in Russian*)
[4] Babkin A V, Kolpakov V I, Ohitin V N 2006 *Numerical methods in problems of physics of fast-flowing processes* (Moscow: Bauman Moscow State Technical University publishing) (*in Russian*)
[5] Hallquist J O 2019 *LS-DYNA Theory Manual* (Livermore: Livermore Software Technology Corporation)

[6] Melvin J, Weber K 1985 *Task B Final Report* UMTRI-85-3 University of Michigan Transportation Research Institute, Ann Arbor, MI.
[7] Olovsson L 2007 *Corpuscular method for airbag deployment simulations in LS-DYNA* Report R32S-1 IMPETUSafea AB.
[8] Lobdell T E, Kroell C K, Schneider D C, Hering W E and Nahum A M 1973 *Human Impact Response: Measurement and Simulation* King W F and Mertz H J (eds.) (Plenum Press, New York) 201-245
[9] Horsch J, Patrick L 1976 Cadaver and Dummy knee impact response, SAE 760799, 20th Stapp Car Crash Conference, Warrendale, PA.

IOP Conf. Series: Materials Science and Engineering 1191 (2021) 012004 doi:10.1088/1757-899X/1191/1/012004

Matrix method for constructing dynamical equations of snowboarder

Borisov A.V.[1], Kaspirovich I.E.[2], Mukharlyamov R.G.[2]

[1]Branch OF National Research University "Moscow Power Engineering Institute" in Smolensk, Energeticheskiy proezd, 1, Smolensk, 214013, Russia
[2]Institute of Physical Research and Technologies, Peoples' Friendship University of Russia, Miklukho-Maklai str., 6, Moscow, 117198, Russia

E-mail: BorisowAndrej@yandex.ru

Abstract. *A method for setting equations of motion for a system of solids with nonholonomic constraints is proposed. The construction of dynamical equations involves the application of a matrix form of recording. Based on the analysis of systems with one, two, or three links, generalizations are derived for the matrices contained in the matrix method for constructing a system with an arbitrary finite number of links. An algorithm for describing the model of a snowboarder with an arbitrary finite number of links of a complex structure with variable length, sliding on an absolutely solid ski, is compiled. A non-holonomic constraint occurs in the ski-snow contact zone. The problem of constructing a stable motion of a system with variable-length links using the link stabilization method is considered. The control is implemented by changing the length of the links and changing the angles between the links. The connection of the links is modeled by two cylindrical joints, which provide significant mechanical strength, simplicity of design and location of the control drives.*

1. Introduction

The problem of developing effective methods for composing systems of differential equations of motion for anthropomorphic mechanisms with a large number of links is relevant due to the cumbersome nature of such systems, which require significant resources and time. Methods for setting systems of differential equations of motion of an anthropomorphic mechanism with holonomic constraints based on matrix and recurrent methods are described in [1-3]. The arguments in favor of the fact that when modeling the human musculoskeletal system, the links of the kinematic chain change their length and cannot be considered absolutely solid, are described in works [4-7]. The papers [8-14] is devoted to the problem of modeling the dynamics of the mechanism with regard for nonholonomic constraints. It offers a method for solving the problem of controlling the dynamics of a system with program constraints. The proposed model of a variable-length link finds, for example, application in modeling exoskeletons [1-3], because the available models are currently uncomfortable and have a short battery life and are produced in single copies [15-31].

In this paper is devoted to the application of the matrix method for constructing a system of differential equations of motion for a snowboarder model with a finite number of links of variable length.

IOP Conf. Series: Materials Science and Engineering 1191 (2021) 012004 doi:10.1088/1757-899X/1191/1/012004

2 Description of the model of a skier-snowboarder with links of variable length on skis

A model of a snowboarder consisting of k movable links of variable length, pivotally fixed on a completely solid inertial ski (Fig. 1). The rotation of the links occurs at joints, which are attached at two points to maintain the link in a vertical plane and turns the link in the plane of motion. There is a longitudinal force acting along the rod and providing a change in its length. There are arbitrary number of concentrated masses α_i $(i = 1,2,\dots,k)$ possessed at i-th rod. The masses are considered to be concentrated: at points: $A_i = C_{i1}$ we have a mass m_{i1}, at point C_{i2} mass m_{i2}, …, at point $C_{i\alpha i}$ mass $m_{i\alpha i}$, $i = 1,2,\dots,k$. The position of the masses on the rod is given by constant multipliers n_{i1}, n_{i2}, …, $n_{i\alpha 1}$ respectively $(0 \le n_{i\beta} \le 1, \beta = 1,2,\dots,\alpha_i, i = 1,2,\dots,k)$. The first index of the multiplier indicates the number of the link, the second-the number of the mass on the link. Let the ski has coordinates x_{A1}, y_{A1} and a rotation angle φ_0, and it moves in the plane XOY. The position of the link A_1A_2 is determined by two angles: φ_1 – angle between axis OX and a projection of the link A_1A_2 to the plane XOY, measured from the OX axis counterclockwise; ψ_1 – the angle between the link A_1A_2 and its projection on the XOY plane, calculated from the projection of the link A_1A_2 on the XOY plane counterclockwise; the change in its length l_1 and the coordinates of the pole $A1(x_{A1}, y_{A1})$. The position of the link A_2A_3 is determined by two angles: φ_1, ψ_1 and the change in its length l_2. The position of the other links is determined in the same way.

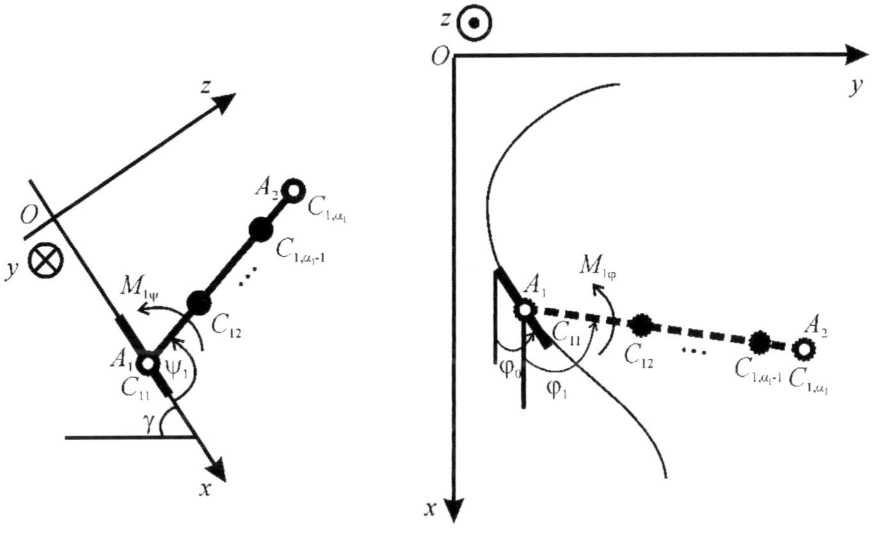

Figure 1. Model of a snowboarder in the form of a single movable link of variable length on skis, sliding along an inclined plane.

The ski simulates a snowboard and two athlete's feet rigidly fixed on it. The A_1 A_2 link models both legs of the skier, link A_2A_3 corresponds to two thighs. This choice of the lower limb model is due to the fact that both legs of the athlete are rigidly fixed to the skis and run parallel to the pelvic-femoral joint.

The torque that is applied to the ski and created by the leg controls the movement and is determined by the components $M_{1\psi}$ and $M_{1\varphi}$ acting in the ankle joint. All joints are modeled by combinations of two cylindrical joints that provide the necessary mobility of the model of the human musculoskeletal system.

Nonholonomic constraint along the contact surface of the ski and snow, provides the trajectory of the snowboarder described by a sinusoid. Taking into account the equation of nonholonomic connection over the contact surface of the ski and snow, which imposes a restriction on coordinate variations, using the Routh equations [14], a system of differential equations of motion is obtained (1)-(5).

$$
\begin{aligned}
&(m_0 + \sum_{\beta=1}^{\alpha_1} m_{1\beta})\mathrm{sc}^2\varphi_0[\,\ddot{x} + \mathrm{tg}\varphi_0\,\dot{x}\dot{\varphi}_0\,] + l_1 \sum_{\beta=1}^{\alpha_1} m_{1\beta}n_{1\beta}\,[(\,C_1^{\varphi}\,\mathrm{tg}\varphi_0 - S_1^{\varphi}\,)\,C_1^{\psi}\,\ddot{\varphi}_1 - \\
&-(\,S_1^{\varphi}\,\mathrm{tg}\varphi_0 + C_1^{\varphi}\,)\,S_1^{\psi}\,\ddot{\psi}_1 - (\,S_1^{\varphi}\,\mathrm{tg}\varphi_0 + C_1^{\varphi}\,)\,C_1^{\psi}\,\dot{\varphi}_1^2 - (\,S_1^{\varphi}\,\mathrm{tg}\varphi_0 + C_1^{\varphi}\,)\,C_1^{\psi}\,\dot{\psi}_1^2 - \\
&-2(\,C_1^{\varphi}\,\mathrm{tg}\varphi_0 - S_1^{\varphi}\,)\,S_1^{\psi}\,\dot{\varphi}_1\,\dot{\psi}_1 - 2(\,S_1^{\varphi}\,\mathrm{tg}\varphi_0 + C_1^{\varphi}\,)\,S_1^{\psi}\,\dot{l}_1\,\dot{\psi}_1 - 2(\,C_1^{\varphi}\,\mathrm{tg}\varphi_0 + S_1^{\varphi}\,)\,C_1^{\psi}\,\dot{l}_1\,\dot{\varphi}_1\,] + \\
&+\sum_{\beta=1}^{\alpha_1} m_{1\beta}n_{1\beta}\,(\,S_1^{\varphi}\,\mathrm{tg}\varphi_0 + C_1^{\varphi}\,)\,C_1^{\psi}\,\ddot{l}_1 = (m_0 + \sum_{\beta=1}^{\alpha_1} m_{1\beta}\,)g_1 + R_x + R_y\mathrm{tg}\varphi_0,
\end{aligned}
\tag{1}
$$

$$
\ddot{\varphi}_0 = 0, \tag{2}
$$

$$
\begin{aligned}
&l_1 \sum_{\beta=1}^{\alpha_1} m_{1\beta}n_{1\beta}\,(\,C_1^{\varphi}\,\mathrm{tg}\varphi_0 - S_1^{\varphi}\,)\,C_1^{\psi}\,\ddot{x} + l_1^2 \sum_{\beta=1}^{\alpha_1} m_{1\beta}n_{1\beta}^2\,(\,C_1^{\psi}\,)^2\,\ddot{\varphi}_1 - \\
&-2l_1^2 \sum_{\beta=1}^{\alpha_1} m_{1\beta}n_{1\beta}^2\;S_1^{\psi}\,C_1^{\psi}\,\dot{\varphi}_1\dot{\psi}_1 + 2l_1 \sum_{\beta=1}^{\alpha_1} m_{1\beta}n_{1\beta}^2\,(\,C_1^{\psi}\,)^2\,\dot{\varphi}_1\dot{l}_1 - \\
&-l_1 \sum_{\beta=1}^{\alpha_1} m_{1\beta}n_{1\beta}\;C_1^{\psi}\,C_1^{\varphi}\,\mathrm{sc}^2\varphi_0\,\dot{x}\dot{\varphi}_0 = M_{1\varphi} + R_x l_1 \sin\varphi_1 + R_y l_1 \cos\varphi_1,
\end{aligned}
\tag{3}
$$

$$
\begin{aligned}
&-l_1 \sum_{\beta=1}^{\alpha_1} m_{1\beta}n_{1\beta}\,(\,S_1^{\varphi}\,\mathrm{tg}\varphi_0 + C_1^{\varphi}\,)\,S_1^{\psi}\,\ddot{x} + l_1^2 \sum_{\beta=1}^{\alpha_1} m_{1\beta}n_{1\beta}^2\;\ddot{\psi}_1 + l_1^2 \sum_{\beta=1}^{\alpha_1} m_{1\beta}n_{1\beta}^2\;S_1^{\psi}\,C_1^{\psi}\,\dot{\varphi}_1^2 + \\
&+2l_1 \sum_{\beta=1}^{\alpha_1} m_{1\beta}n_{1\beta}^2\;\dot{\psi}_1\dot{l}_1 - l_1 \sum_{\beta=1}^{\alpha_1} m_{1\beta}n_{1\beta}\;S_1^{\varphi}\,S_1^{\psi}\,\mathrm{sc}^2\varphi_0\,\dot{x}\dot{\varphi}_0 = M_{1\psi} + R_x l_1 \sin\psi_1 + R_z l_1 \cos\psi_1,
\end{aligned}
\tag{4}
$$

$$
\begin{aligned}
&\sum_{\beta=1}^{\alpha_1} m_{1\beta}n_{1\beta}\,(\,S_1^{\varphi}\,\mathrm{tg}\varphi_0 + C_1^{\varphi}\,)\,C_1^{\psi}\,\ddot{x} + \sum_{\beta=1}^{\alpha_1} m_{1\beta}n_{1\beta}^2\;\ddot{l}_1 - l_1 \sum_{\beta=1}^{\alpha_1} m_{1\beta}n_{1\beta}^2\,(\,C_1^{\psi}\,)^2\,\dot{\varphi}_1^2 - \\
&-l_1 \sum_{\beta=1}^{\alpha_1} m_{1\beta}n_{1\beta}^2\;\dot{\psi}_1^2 + \sum_{\beta=1}^{\alpha_1} m_{1\beta}n_{1\beta}\;S_1^{\varphi}\,C_1^{\psi}\,\mathrm{sc}^2\varphi_0\,\dot{x}\dot{\varphi}_0 = F_1.
\end{aligned}
\tag{5}
$$

Further, in a similar way, a system of differential equations of motion was compiled for a model with two links (Fig. 2).

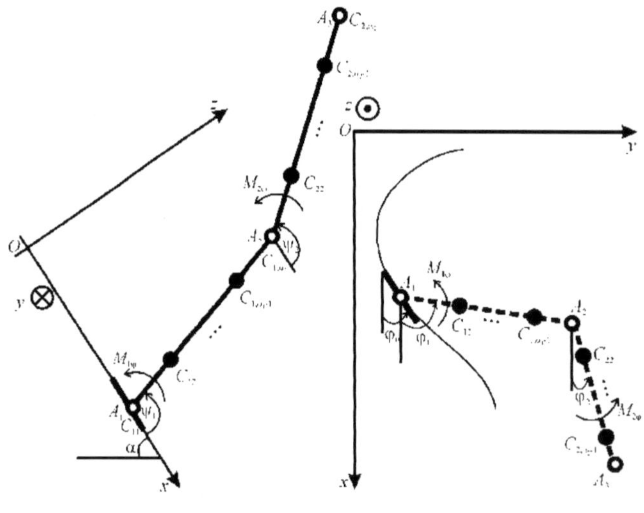

Figure 2. Model of a snowboarder in the form of two movable links of variable length on skis, sliding along an inclined plane.

Because the resulting system of differential equations of motion for the model of a skier-snowboarder with two movable links of variable length turned out to be cumbersome, we will give only the first equation (6).

$$(m_0 + \sum_{\beta=1}^{\alpha_1} m_{1\beta} + \sum_{\beta=1}^{\alpha_2} m_{2\beta})\mathrm{sc}^2\varphi_0[\ddot{x} + \mathrm{tg}\varphi_0\, \dot{x}\dot{\varphi}_0] + l_1(\sum_{\beta=1}^{\alpha_1} m_{1\beta}n_{1\beta} + \sum_{\beta=1}^{\alpha_2} m_{2\beta})[(C_1^{\varphi}\mathrm{tg}\varphi_0 - S_1^{\varphi})C_1^{\psi}\ddot{\varphi}_1 -$$

$$-(S_1^{\varphi}\mathrm{tg}\varphi_0 + C_1^{\varphi})S_1^{\psi}\ddot{\psi}_1 - (S_1^{\varphi}\mathrm{tg}\varphi_0 + C_1^{\varphi})C_1^{\psi}\dot{\varphi}_1^2 - (S_1^{\varphi}\mathrm{tg}\varphi_0 + C_1^{\varphi})C_1^{\psi}\dot{\psi}_1^2 - 2(C_1^{\varphi}\mathrm{tg}\varphi_0 - S_1^{\varphi})S_1^{\psi}\dot{\varphi}_1\dot{\psi}_1 -$$

$$-2(S_1^{\varphi}\mathrm{tg}\varphi_0 + C_1^{\varphi})S_1^{\psi}\dot{l}_1\dot{\psi}_1 - 2(C_1^{\varphi}\mathrm{tg}\varphi_0 + S_1^{\varphi})C_1^{\psi}\dot{l}_1\dot{\varphi}_1] + (\sum_{\beta=1}^{\alpha_1} m_{1\beta}n_{1\beta} + \sum_{\beta=1}^{\alpha_2} m_{2\beta})(S_1^{\varphi}\mathrm{tg}\varphi_0 + C_1^{\varphi})C_1^{\psi}\ddot{l}_1$$

$$+$$

$$+ l_2(\sum_{\beta=1}^{\alpha_2} m_{2\beta}n_{2\beta})[(C_2^{\varphi}\mathrm{tg}\varphi_0 - S_2^{\varphi})C_2^{\psi}\ddot{\varphi}_2 - (S_2^{\varphi}\mathrm{tg}\varphi_0 + C_2^{\varphi})S_2^{\psi}\ddot{\psi}_2 - (S_2^{\varphi}\mathrm{tg}\varphi_0 + C_2^{\varphi})C_2^{\psi}\dot{\varphi}_2^2 - \quad (6)$$

$$-(S_2^{\varphi}\mathrm{tg}\varphi_0 + C_2^{\varphi})C_2^{\psi}\dot{\psi}_2^2 - 2(C_2^{\varphi}\mathrm{tg}\varphi_0 - S_2^{\varphi})S_2^{\psi}\dot{\varphi}_2\dot{\psi}_2 - 2(S_2^{\varphi}\mathrm{tg}\varphi_0 + C_2^{\varphi})S_2^{\psi}\dot{l}_2\dot{\psi}_2 -$$

$$-2(C_2^{\varphi}\mathrm{tg}\varphi_0 + S_2^{\varphi})C_2^{\psi}\dot{l}_2\dot{\varphi}_2] + \sum_{\beta=1}^{\alpha_2} m_{2\beta}n_{2\beta}(S_2^{\varphi}\mathrm{tg}\varphi_0 + C_2^{\varphi})C_2^{\psi}\ddot{l}_2 =$$

$$= (m_0 + \sum_{\beta=1}^{\alpha_1} m_{1\beta} + \sum_{\beta=1}^{\alpha_2} m_{2\beta})g_1 + R_x + R_y\mathrm{tg}\varphi_0,$$

Considering for generalization the model of a skier-snowboarder with n movable links of variable length (Fig. 3) on the basis of models with one and two links, the required matrix form of recording the system of differential equations of motion was obtained (7).

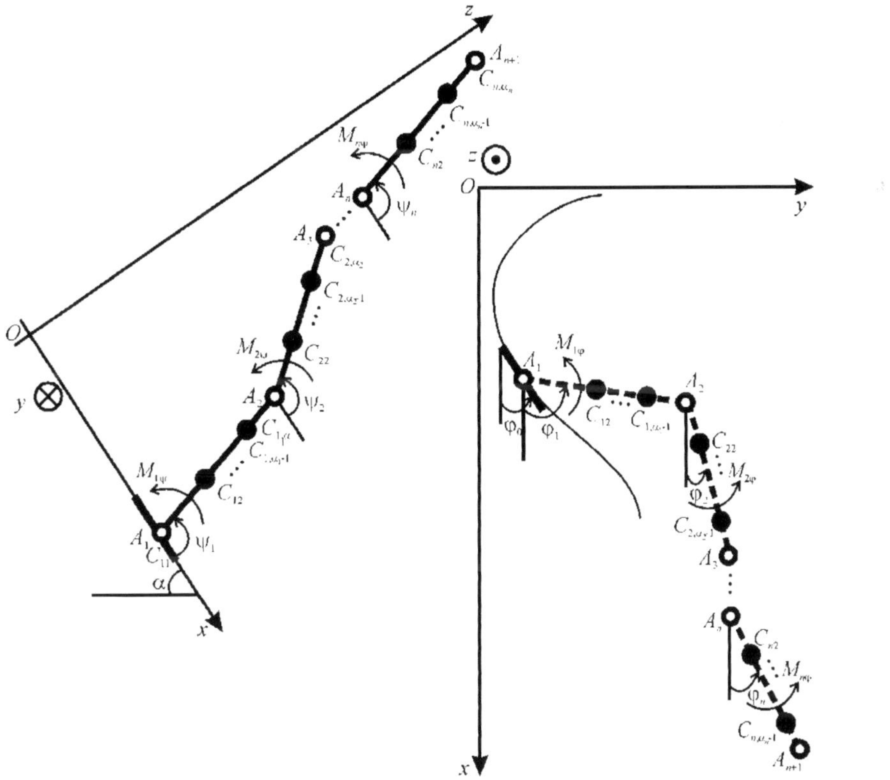

Figure 3. Model of a snowboarder in the form of n movable links of variable length on skis, sliding along an inclined plane.

$$\begin{gathered} A_\kappa(\varphi,\psi,l)\ddot{\varphi} + B_\kappa(\varphi,\psi,l)\ddot{\psi} + C_\kappa(\varphi,\psi,l)\ddot{x} + \Lambda_\kappa(\varphi,\psi,l)\ddot{l} + \\ + D_\kappa(\varphi,\psi,l)\dot{\Phi}\dot{\varphi} + E_\kappa(\varphi,\psi,l)\dot{\Psi}\dot{\psi} + 2G_\kappa(\varphi,\psi,l)\dot{\Phi}\dot{\psi} + 2H_\kappa(\varphi,\psi,l)\dot{\Phi}\dot{l} + \\ + 2K_\kappa(\varphi,\psi,l)\dot{\Psi}\dot{l} + N_\kappa(\varphi,\psi,l)\dot{x}\dot{\Phi} + gP_\kappa(\psi) = M_\kappa(\varphi,\psi,l), \end{gathered} \tag{7}$$

where lower indices of the matrices indicate the description of the corresponding generalized coordinate: $\kappa = 1, 2, 3$, where 1 corresponds to the generalized coordinate φ, $2 - \psi$, $3 - x$, $4 - l$; where: φ, ψ – angular generalized coordinates $\varphi = (\varphi_1,\ldots,\varphi_n)^T$ и $\psi = (\psi_1, \ldots, \psi_n)^T$ – angular vectors; x – linear generalized coordinate; l - generalized coordinates associated with variations in link lengths $l = (l_1, \ldots, l_n)^T$ – variable-length link length vector; $\dot{\varphi}$, $\dot{\psi}$ – angular velocity vectors; $\ddot{\varphi}$, $\ddot{\psi}$ – the vectors of angular accelerations; $\dot{\Phi} = \mathrm{diag}(\dot{\varphi}_1,\ldots,\dot{\varphi}_n)$, $\dot{\Psi} = \mathrm{diag}(\dot{\psi}_1,\ldots,\dot{\psi}_n)$ – diagonal matrices; M_κ – vectors of generalized forces; $A_\kappa(\varphi,\psi,l)$, $B_\kappa(\varphi,\psi,l)$, $D_\kappa(\varphi,\psi,l)$, $E_\kappa(\varphi,\psi,l)$, $G_\kappa(\varphi,\psi,l)$, $N_\kappa(\varphi,\psi,l)$ – matrices that take into account inertial properties; $\Lambda_\kappa(\varphi,\psi,l)$, $H_\kappa(\varphi,\psi,l)$, $K_\kappa(\varphi,\psi,l)$ – matrices that take into account the variable length of links; $P_\kappa(\psi)$ – matrices defined by the moments of gravity forces.

Expressions of elements of the matrix contained in the system of differential equations of motion of a mechanical system obtained from the analysis of models with one, two links of variable length are determined. The matrix A_φ is symmetrical with components a_{ij}^{φ} (8), if i – number of line, j – number of row, $i, j = 1,2, \ldots, 8$, $j \geq i$:

$$a_{ij}^{\varphi} = (\delta_{ij}\zeta_i + (1 - \delta_{ij})\eta_j + \sum_{k=j+1}^{8}\theta_k)l_il_j\cos(\varphi_i - \varphi_j)\cos\psi_i\cos\psi_j, \tag{8}$$

where: δ_{ij} – Kronecker delta: $\delta_{ij} = \begin{cases}1, & i = j, \\ 0, & i \neq j.\end{cases}$, $\zeta_i = \sum_{\beta=1}^{\alpha_i} m_{i\beta}n_{i\beta}^2$, $\eta_i = \sum_{\beta=1}^{\alpha_i} m_{i\beta}n_{i\beta}$, $\theta_i = \sum_{\beta=1}^{\alpha_i} m_{i\beta}$, α_i – the number of concentrated masses on the i-th link.

3. Results and discussion

Then, a numerical simulation of the motion of a snowboarder with a given number of links was performed using the link stabilization method [8-14]. As a result of numerical simulation of the movement of a skier-snowboarder with two links of variable length, which have their values close to the average for a person of ordinary constitution for the lower leg and thigh [1-5,14], the solution of the Cauchy problem is obtained. Let us give as an illustration a graph of the deviation from the connection [8-14] with a non-zero initial deviation (Fig. 4).

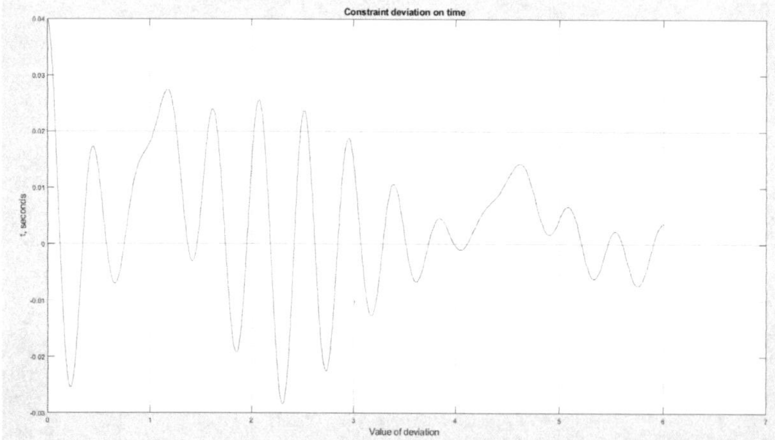

Figure 4. Deviation from constraint with nonzero initial deviation for ψ_1.

The dependence of the connection deviation presented in Figure 4 clearly demonstrates the stability and efficiency of the used method of stabilizing the connection for numerical modeling of the movements of a skier-snowboarder with links of variable length by solving the Cauchy problem for a system of differential equations of motion of the form (6).

4. Conclusions

Thus, a matrix method for setting a system of differential equations for a snowboarder that simulates real human movements has been developed. A generalized model with any finite number of links was created, which allows modeling the dynamics of a skier-snowboarder with a given accuracy.

A numerical calculation of the movements of a model of a skier-snowboarder containing two links of variable length on an absolutely rigid ski was carried out using biomechanical and anatomical data. The effectiveness of the applied method of stabilization of bonds is established.

The model of a snowboarder can be useful in various areas of human activity. For example, in the training process in sports for practicing complex coordination movements of highly qualified skiers-snowboarders, when creating anthropomorphic robots and exoskeletons.

Acknowledgments: The work was performed as financial support from RFBR grant under project No. 19-08-00261 A.

References

[1] Borisov A V, Rozenblat G M (2017) Matrix method of constructing the differential equations of motion of an exoskeleton and its control // *Journal of Applied Mathematics and Mechanics* **81**, pp 351-359.

[2] Borisov A V, Rozenblat G M (2018) Modeling the dynamics of an exoskeleton with control torques in the joints and a variable length of the links using the recurrent method for constructing differential equations of motion // ISSN 1064-2307, *Journal of Computer and Systems Sciences International*, vol. **57**, No. **2**, pp 319-347.

[3] Borisov A V (2019) Two-dimensional and three-dimensional models of anthropomorphic robot and exoskeleton with links of variable length. // *Proc. of 24th Int. Conf. "MECHANIKA 2019"*. 17 May 2019 Kaunas University of Technology, Lithuania. pp 26-39.

[4] Borisov A. V., Chigarev A.V. (2020) The Causes of a Change in The Length of a Person's Link and Their Consideration When Creating an Exoskeleton // *Biomedical Journal of Scientific and Technical Research.* ISSN: 2574 -1241. – 2020. – Volume 25. – Issue 1. – P. 18769-18771. https://biomedres.us/volume25-issue1.php https://biomedres.us/pdfs/BJSTR.MS.ID.004137.pdf

[5] Borisov A. V., Chigarev A.V. (2011) Simulation of controlled motion of the bipedal anthropomorphic mechanism // *Russian Journal of Biomechanics.* 2011. vol. **15**, No. **1 (51)**, pp. 69-83.

[6] Piña-Martínez E., Rodriguez-Leal E. Inverse Modeling of Human Knee Joint Based on Geometry and Vision Systems for Exoskeleton Applications // *Mathematical Problems in Engineering.* Vol. **2015.** Article ID 145734, **14** pages, 2015. https://doi.org/10.1155/2015/145734. Режим доступа: http://www.hindawi.com/journals/mpe/2015/145734/

[7] Basic biomechanics of the musculoskeletal system / [edited by] Margareta Nordin, Victor H. Frankel. – 3rd ed. – Lippincott Williams & Wilkins, 2001. – **467** P.

[8] Baumgarte J. (1972) Stabilization of Constraints and Integrals of Motion in Dynamical Systems // Computer Methods in Applied Mechanics and Engineering. 1972. V. **1.** Iss. **1.** P. 1-16.

[9] Mukharlyamov R.G., Ibragimov R.G., Kolesnikov A.P. (1990) Control in Adaptive Optical

Systems // *IFAC Workshop on Evaluation of Adaptive Control Strategies in Industrial Application. Tbilisi, USSR*, 17-21 Oct. 1989. Oxford : Pergamonpress, 1990. P. 301-306.

[10] Mukharlyamov R.G. (2007) Reduction of Dynamical Equations for the Systems with Constraints to Given Structure // *J. Applied Mathematics and Mechanics*. 2007. V. **71.** № **3**. P. 401-410.

[11] Mukharlyamov R.G., Beshaw A.W. (2013) Solving Differential Equations of Motion for Constrained Mechanical Systems // *RUDN Bulletin, Series. Mathematics, computer science, physics*. 2013, № **3**. P. 81-91.

[12] Mukharlyamov R.G., Tleubergenov M.I. (2017) Control of System Dynamics and Constraints Stabilization *// Distributed Computer and Communication Networks (DCCN 2017) / Eds V. Vishnevskiy, K. Samouylov, Kozyrev D. Communications in Computer and Information Science*. V. **700**. Cham : Springer. P. 431-442.

[13] Kaspirovich I.E. (2016) Application of Constraint Stabilization to Nonholonomic mechanics // *2nd Intern. Conf. on Industrial Engineering, Applications and Manufacturing (ICIEAM)*, Chelyabinsk, 2016. DOI 10.1109/ICIEAM.2016.7910921.

[14] Andrey V. Borisov, Ivan E.Kaspirovich, Robert G. Mukharlyamov. (2019) Dynamics control of nonholonomic system modelling motion of skier and snowboarder. // *ANS Conference Series: Scientific Heritage of Sergey A.Chaplygin (Nonholonomic Mechanics, Vortex Structures and Hydrodynamics) : Book of Abstracts.* — Moscow–Izhevsk : Institute of Computer Science, 2019. – P. 37-39.

[15] Tsukahara A., Hasegawa Y., Sankai Y. (2011) Gait Support for Complete Spinal Cord Injury Patient by Synchronized Leg-swing with HAL // *Proc. IEEE/RSJ Intern. Conf. on Intelligent Robots and Systems (IROS '11)*. San Francisco, 2011. P. 1737-1742.

[16] Tsukahara A., Hasegawa Y., Eguchi K., Sankai Y. (2015) Restoration of Gait for Spinal Cord Injury Patients Using HAL with Intention Estimator for Preferable Swing Speed // IEEE Transactions on Neural Systems and Rehabilitation Engineering. 2015. V. **23**. №. **2.** P. 308-318.

[17] Hassan M., Kadone H., Suzuki K., Sankai Y. (2012) Exoskeleton Robot Control Based on Cane and Body Joint Synergies // Proc. 25th IEEE/RSJ Intern. Conf. on Robotics and Intelligent Systems (IROS '12). Vilamoura, 2012. P. 1609-1614.

[18] Kazerooni H. (2005) Exoskeletons for Human Power Augmentation *// Proc. IEEE IRS/RSJ Intern. Conf. on Intelligent Robots and Systems (IROS '05).* Edmonton, 2005. P. 3120-3125.

[19] Bortole M., A. del Ama, Rocon E., Moreno J.C., Brunetti F., Pons J.L. (2013) A Robotic Exoskeleton for Overground gait Rehabilitation *// Proc. of the IEEE Intern. Conf. on Robotics and Automation (ICRA '13).* Karlsruhe, 2013. P. 3356-3361.

[20] Qing Guo, Songjing Li, Dan Jiang. (2015) A Lower Extremity Exoskeleton: Human-Machine Coupled Modeling, Robust Control Design, Simulation, and Overload-Carrying Experiment // *Mathematical Problems in Engineering*, Vol. 2015, pp. 1-15; doi:10.1155/2015/905761.

[21] Pavel Blažek, Josef Bydžovský, Robert Griffin, Karel Mls, Brandon Peterson. (2020) Obstacle Awareness Subsystem for Higher Exoskeleton Safety. *// World Symposium on Digital Intelligence for Systems and Machines DISA* 2020: Towards Digital Intelligence Society P. 59-71.

[22] W. Bao, D. Villarreal, J. -C. Chiao. (2020) Vision-Based Autonomous Walking in a Lower-Limb Powered Exoskeleton. *// 2020 IEEE 20th International Conference on Bioinformatics and Bioengineering (BIBE),* Cincinnati, OH, 2020, P. 830-834, doi: 10.1109 BIBE50027.2020.00141.

[23] Proud JK, Lai DTH, Mudie KL, Carstairs GL, Billing DC, Garofolini A, Begg RK. (2020) Exoskeleton Application to Military Manual Handling Tasks. *Hum Factors.* 2020 Nov 18:18720820957467. doi: 10.1177/0018720820957467. Epub ahead of print. PMID: 33203237.

[24] Passive Exoskeleton: patent № US7571839B2 United States / Conrad Chu, Piscataway, NJ

(US), Andy Chu, Cambridge, MA (US), publication of 21.07.2009.
[25] Projects. ExoChair. Industrial equipment. Retrieved from: https://karfidovlab.com/projects/exochair/
[26] Exoatlet. Exoskeleton for medical and social rehabilitation. Retrieved from: https://exoatlet.ru/
[27] Meet the indego exoskeleton. Retrieved from: http://www.indego.com/indego/us/en/home
[28] Powered Exoskeleton Published Research Summary. Retrieved from: https://rewalk.com/
[29] The leader in exoskeleton technology. Retrieved from: https://eksobionics.com/
[30] Raytheon Unveils Lighter, Faster, Stronger Second Generation Exoskeleton Robotic Suit. Retrieved from: http://raytheon.mediaroom.com/index.php?s=43&item=1652
[31] US Army trials exoskeletons for military use. Retrieved from: https://www.army-technology.com/features/us-army-exoskeletons/

Surface gravity waves in a wedge-shaped vessel

V A Kalinichenko[1,2]

[1]Ishlinsky Institute for Problems in Mechanics of the Russian Academy of Sciences, Moscow, 119526 Russian Federation
[2]Theoretical Mechanics Department, Bauman Moscow State Technical University, Moscow, 105005 Russian Federation
E-mail: kalin@ipmnet.ru vakalin@mail.ru

Abstract. The results of experiments on the excitation of standing surface gravity waves in a cylindrical vessel with a wedge-shaped base are presented. The fundamental difference from the liquid wave motions in a rectangular vessel is the absence of spatial symmetry for the profile of maximum development and an increase in the wave amplitude towards the wedge tip. To describe waves in a wedge-shaped vessel, a numerical-analytical model of long waves in a vessel of variable width and a linear model of standing waves in a vessel with a base in the form of a circular sector with a small central angle are used.

1. Introduction

In hydrodynamics, wave motions of a fluid at parametric resonance are of not only applied but also fundamental interest. A historically complete review of the available theoretical and experimental works on the comprehensive study of Faraday waves in vessels of the simplest shape with rectangular and circular bases is given in [1]. At present, interest in Faraday waves is largely due to the possibility of studying wave motions of a liquid from the standpoint of the general theory of waves. It should be noted the special importance of parametric resonance for hydrodynamic experiments, which consists in the absence of any wavemakers; simplification and clarification of the processes of observation and measurement of wave parameters; excitation of wave modes with predetermined wave height and frequency; and a high degree of repeatability of experimental conditions. A number of effects inherent in any standing gravity waves are experimentally studied using Faraday waves. For example, parametric resonance is used to study the mechanism of breaking of standing waves on the free surface of a liquid in a circular cylinder [2], and the search for ways to regularize the breaking waves is used in a rectangular vessel [3].

In present work, under the conditions of Faraday resonance, a special mode of fluid oscillations in a narrow wedge-shaped vessel is investigated. The specified geometry of the vessel was previously used in [4], however, no quantitative estimates were made. The problem of standing waves in a cylinder with a base in the form of a wedge (pointed elongated triangle) is similar in formulation to waves in cylindrical tanks for which the base is a sector of a circle with a small central angle. In the 1960s, to reduce the intense vibrations of rocket fuel in cylindrical tanks with a circular base, it was proposed to install diametrical partitions, which would provide a transition to tanks with a base in the form of a circular sector – compartmented cylindrical tanks. The natural frequency of fluid oscillations in such tanks increases, while the structural dynamic loads are reduced. A number of experiments have been devoted to assessing the frequency and force factors; see, for example, [5, 6]. Note that tanks with a sector angle below $\pi/4$ were not used in the experiments.

In this work, we experimentally investigate standing gravity Faraday waves in a cylindrical vessel with a wedge-shaped base, the apex angle of which is $\pi/36$. A comparison is made with waves in a rectangular cylinder. This problem may have practical application to the phenomenon of seiches [7] or tides in strongly convergent channels [8].

2. Experimental procedure

We studied the third and fourth wave modes (n = 3, 4) in cylindrical vessels with a rectangular base 60×5.2 cm (figure 1a) and with a base in the form of a wedge or a pointed elongated right triangle with legs $l = 60$ and $b = 5.2$ cm and apex angle $\alpha = \pi/36 = 5^0$ (figures1 b – d). The vessels were filled with water used as a working fluid to a depth of $h = 3.8 - 4$ cm – Figure 1d. We note a significant capillary rise of water at the top of the wedge (figure 1d). This effect [9] is characteristic of a liquid between two vertical planes located at a small angle, but for our experiments it is parasitic, since it leads to a local distortion of the wave profile.

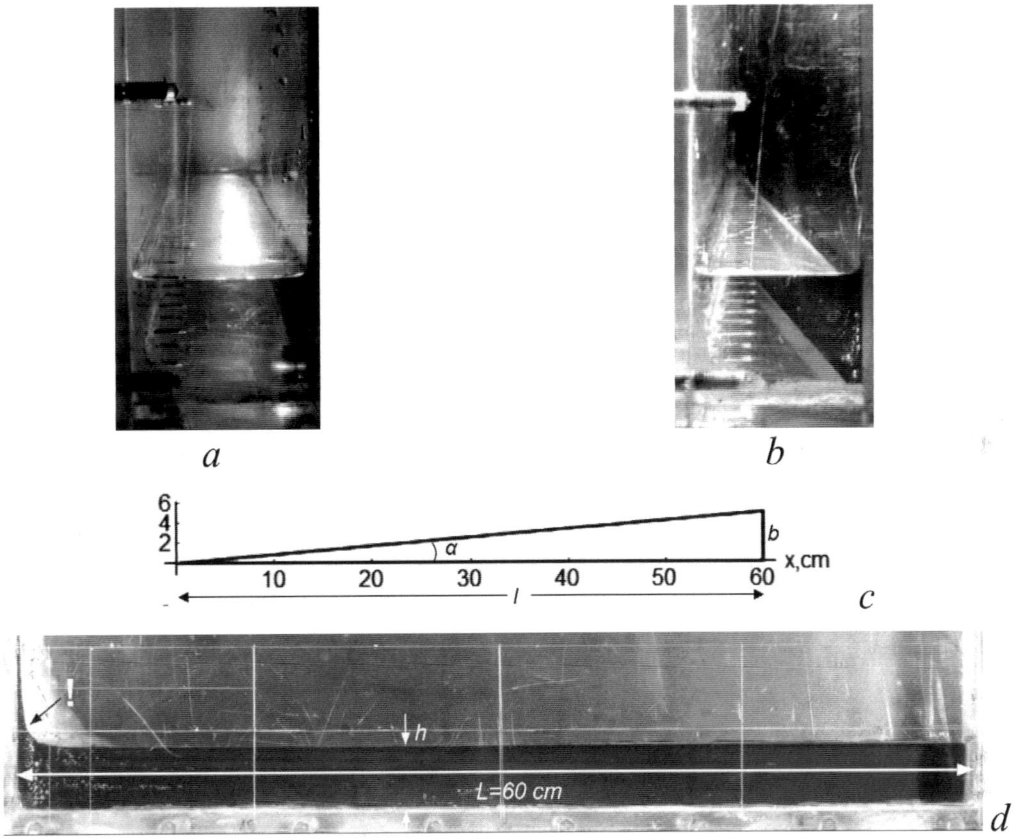

Figure 1. (a) Cylinder with rectangular base 60×5.2 cm (view from the end of the vessel). (b – d) Cylindrical vessel with a wedge-shaped base – end view, plan view and frontal view. In figure d: ! - capillary rise of water at the corner point (d)

The vessel was placed on the platform of an electromechanical shaking table providing harmonic oscillations in a vertical direction. Two-dimensional wave motions were investigated in the regime of the main Faraday resonance [3], when the vessel oscillation frequency Ω is twice the frequency ω of excited waves ($\Omega \sim 2\omega$). With fixed vessel amplitude $s = 0.7$ cm the variations of Ω provided the excitation of the corresponding wave mode of number n and wave height H.

The wave pattern was registered with digital cameras DIMAGE Z2 and Canon PowerShot SX50HS (frame rates were 30 and 120 fps) in a moving reference system rigidly connected with the vessel. The resolution of the video images was 0.15 mm/pixel. The subsequent processing of video frames was carried out by the software ImageJ. All experiments were conducted at room temperature 21–22°C.

3. Results and discussion

Figure 2 shows the profiles of the third (a) and fourth (b) wave modes excited in the wedge-shaped cylinder. The waves had periods $T_{3,4} = 0.645$ and 0.540 s, the corresponding frequencies

$\omega_{3,4} = 2\pi / T_{3,4} = 9.593$ и 11.635 s^{-1}. The video frames in the figure show the dynamics of the wave profiles during one period.

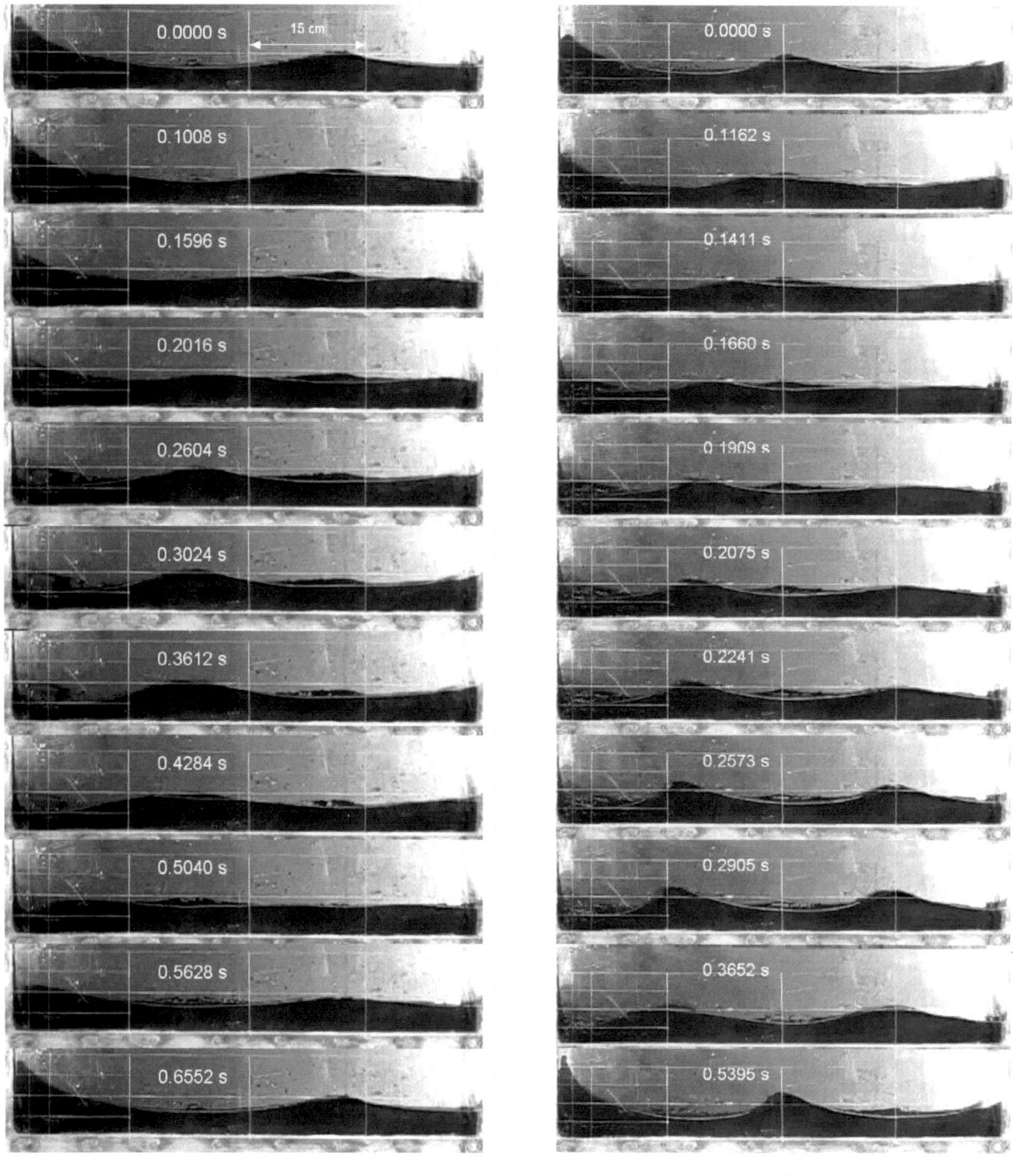

Figure 2. Standing gravitational waves on a free surface of water in a wedge-shaped cylindrical vessel. (a) Third wave mode $n = 3$. (b) Fourth wave mode $n = 4$. (based on video, 120 fps)

First of all, the waves under consideration can be classified as standing, since their antinodes do not move horizontally. The waves are nonlinear, which also manifests itself in the oscillations of the nodes of the standing wave. The node as a fixed point of the wave surface does not exist. In this case, the node can be defined as a point of the free surface that has a zero displacement relative to the unperturbed initial liquid level. The spatial symmetry of the profile of maximum development at $t = 0$ is violated for both modes (figure 2) relative to the vertical drawn through the wave top. The ordinates of the tops of the wave modes at this moment increase from the right to the left wall of the vessel, on which the displacement of the free surface is maximum. Waves are "regular" – they are characterized by temporal periodicity.

The results of experiments with standing waves in a rectangular vessel are shown in figure 3. From the photos (*a, b*) of the wave profiles of maximum development it follows that: the antinodes of the waves do not change their position; nodes make small horizontal vibrations; the ordinates of the wave tops are constant along the vessel. Thus, we have a standing regular nonlinear wave of constant height H.

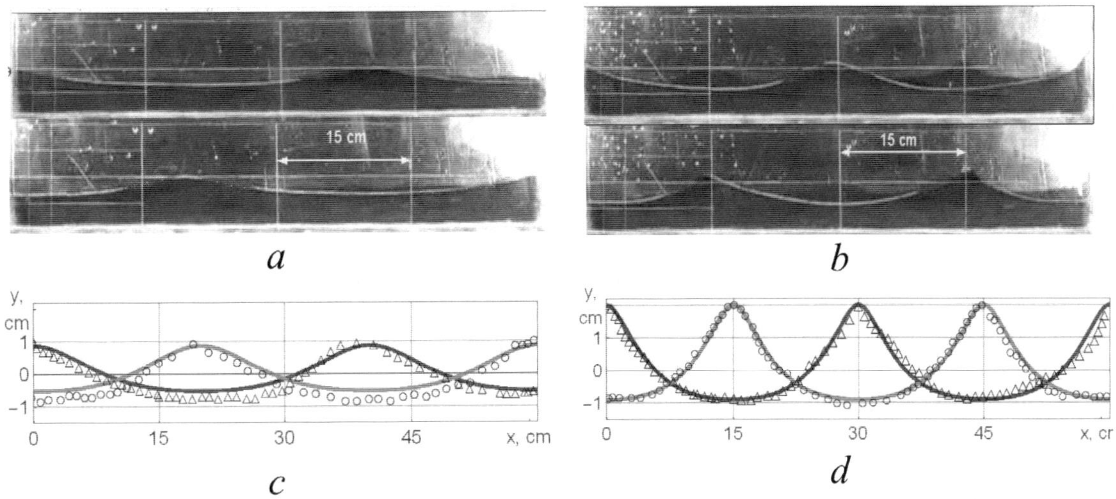

Figure 3. Profiles of waves of maximum development for the third (*a*) and fourth (*b*) modes – $n = 3$ and 4, respectively. (c, d) Comparison of experimental (points) and calculated (curves) profiles. Here: periods $T_{3,4} = 0.706$ and 0.590 s; frequencies $\omega_{3,4} = 2\pi / T_{3,4} = 8.900$ and 10.650 s-1; wave height $H_{3,4} = 1.4$ and 2.9 cm

Using the model [10, 11] of the nonlinear surface Faraday waves in rectangular vessels, it is possible to construct the profile of the free surface ($b = 0$) in the Lagrange variables (a, b, t):

$$\begin{cases} x = a - H\dfrac{\mathrm{ch}kh}{2\mathrm{sh}\kappa h}\sin ka\cos\psi + H^2 k\dfrac{\sin 2ka}{32\mathrm{sh}^2 kh}(1+\cos 2\psi) - \dfrac{3}{64}H^2 k\dfrac{\mathrm{ch}2kh}{\mathrm{sh}^4 kh}\sin 2ka\cos 2\psi \\ \quad + \dfrac{1}{16}H^2 k\dfrac{\mathrm{ch}2kh}{\mathrm{sh}^2 2kh}\sin 2ka, \\ y = \dfrac{H}{2}\cos ka\cos\psi + H^2 k\dfrac{\mathrm{sh}\,2kh}{32\mathrm{sh}^2 kh}(1+\cos 2\psi) + \dfrac{3}{64}H^2 k\dfrac{\mathrm{sh}2kh}{\mathrm{sh}^4 kh}\cos 2ka\cos 2\psi \\ \quad - \dfrac{1}{16}H^2 k\dfrac{1}{\mathrm{sh}^2 2kh}\cos 2ka, \\ \psi = \omega t / 2,\ \lambda = \pi n / l,\ k = 2\pi / \lambda,\ a \in [0, l],\ b \in [-h, 0] \end{cases} \tag{1}$$

The calculated according to (1) and experimental wave profiles are shown in figure 3*c, d*. Comparison of theory and experiment shows that the standing waves observed in experiments are adequately described within the framework of available nonlinear models.

To interpret the results of the present experiment with standing waves in a vessel with a narrow wedge-shaped base, we use the long wave approximation and consider the problem of natural oscillations of an ideal incompressible fluid in an extended channel of variable rectangular cross section [12]. The equation of wave motions of liquid has the form [12]

$$\frac{\partial^2 \eta}{\partial t^2} = \frac{g}{d(x)} \frac{\partial}{\partial x} (S(x) \frac{\partial \eta}{\partial x}) \tag{2}$$

Here, $\eta = \eta(x,t)$ is the fluid elevation, ρ is the fluid density, y is the vertical coordinate, $d(x)$ is the variable vessel width, g is acceleration of gravity. It is assumed that the vessel's variable rectangular cross-section is perpendicular to the x axis (figure 1,c) and its area $S = S(x) = d(x)h$, $0 \le x \le l$, $h = h_0 = \text{const}$ is the fluid depth. For standing waves, the ends of the vessel are limited by rigid vertical walls, on which the boundary impermeability conditions are satisfied

$$\left(\frac{\partial \eta}{\partial x}\right)_{x=0} = \left(\frac{\partial \eta}{\partial x}\right)_{x=l} = 0, \quad t \ge t_o \tag{3}$$

We are looking for periodic solutions to the boundary value problem in the form of $\eta(x,t) = W(x)e^{i\omega t}$, after substitution of which in (2), (3) we obtain

$$\frac{d}{dx}\left(d(x)\frac{dW}{dx}\right) + \frac{d(x)\omega^2}{gh_0} W = 0, \quad W'(0) = W'(l) = 0 \tag{4}$$

If $d(x) = d_0$, then from Eq. (4) we can determine the analytical solution corresponding to the standing wave modes

$$\lambda_n = \frac{\omega^2 l^2}{gh_0} = (\pi n)^2, \; W_n(x) = A\cos\frac{\pi n}{l} x \quad (n = 1,2,3,...)$$

To solve the problem in the case of a variable vessel width, we use a high-precision numerical-analytical method of accelerated convergence [13]. This method was successfully tested when describing the experiment with standing waves in a rectangular tank with local wall and bottom irregularities [14].

The boundary value problem for eigenvalues λ_n and eigenfunctions $W_n(x)$ is reduced to the form

$$\frac{d}{dx}\left(d(x)\frac{dW}{dx}\right) + d(x)\lambda_n W = 0, \quad W'(0) = W'(1) = 0 \tag{5}$$

Here x and d are the horizontal coordinate and the vessel width normalized to l and b, respectively. The unknown parameter λ_n is related to frequency ω as follows

$$\lambda_n = \frac{\omega^2 l^2}{gh_0} = (\pi n)^2$$

To determine the eigenvalues λ_n and the functions $W_n(x)$ of problem (5) with the boundary conditions of the von Neumann type (of the second kind) we will apply the Sturm-Liouville theory and the algorithm of accelerated convergence [13].

Note that the solution of the zero mode is associated with $\lambda_0 = 0$, $W_0(x) \equiv \text{const}$, in particular, $W_0(x) \equiv 0$. Subsequent wave modes n=1,2... require a high-precision solution of the boundary value problem for eigenvalues λ_n and eigenfunctions W_n. Below they are constructed in accordance with the special properties of the vessel width used in the experiment and described by the functions $d(x^*) = x^*$ (figure 4*a*) and $d(x^*) = 1-(b-b_0)x^*/b$ (figure 4*c*) where b =5.2 cm and $10^{-4} \le b_0 \le 10^{-2}$ cm. The derivative boundary condition of $W'(1) = 0$ imposes the specified constraint on the minimum value of b_0. Note that the values of b_0 = 0.0001 – 0.01 cm used in the calculations of wave profiles determine the accuracy of specifying the wedge top for our experiment.

The algorithm of the method consists in successive refinement of the value λ_n based on the discrepancy $\varepsilon = 1-\xi$ ($|\varepsilon| \ll 1$) along the abscissa axis for the node ξ of $W_n'(x)$ corresponding to n and the known derivative $\lambda_n'(\xi)$.

The calculation of the eigenvalues λ_n or the frequencies ω_n is accompanied by the construction of the functions $W_n(x)$. These functions satisfy with the required accuracy the boundary conditions for the given function $d(x)$ and the mode number n ([13]).

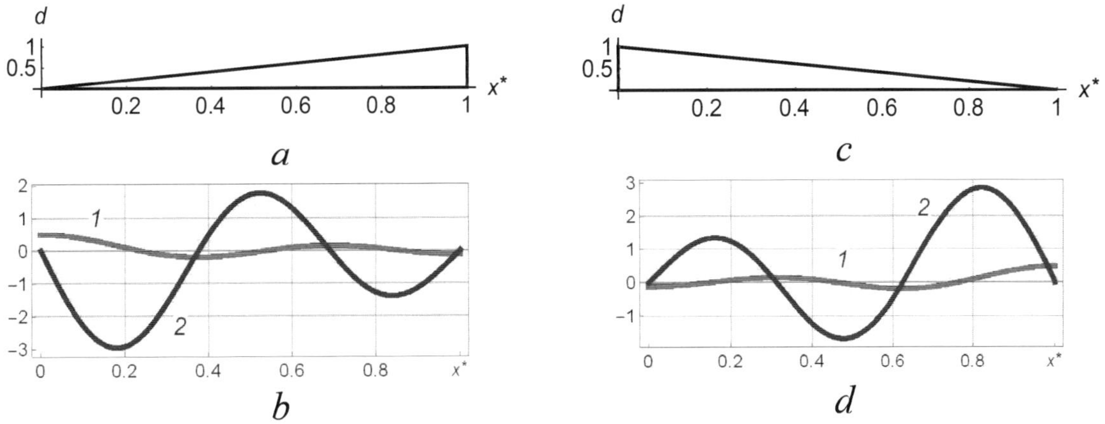

Figure 4. (*b, d*) Eigenfunctions $W_3(x)$ (*1*) and their derivatives $W_3'(x)$ (*2*) for the third wave mode in a cylindrical vessel with base $d(x^*) = x^*$ (*a*) and $d(x^*) = 1-(b-b_0)x^*/b$ (*c*). Here $x^* = x/l$

The need to consider two dependencies $d(x)$ (figure 4*a* and *c*) is due to the fact that for the experiment it is more important to set the displacement of the free surface on the wide end wall of the vessel and numerically estimate the wave height at the wedge top. In the case of $d(x^*) = x^*$, we determine the wave parameters at $x = 0$ (top of the wedge), and the model gives estimates $W(1)$ when the condition $W'(1) = 0$ is met – figure 4*b*. At $d(x^*) = 1-(b-b_0)x^*/b$ (figure 4*c* and *d*), the procedure for numerical determination of $W(1)$ at $W'(1) = 0$ requires the introduction of a small $b_0 \ne 0$. For the purpose of comparison with experiment, for the dimensional dependences $W_n(x)$ obtained in this way, the procedure of spatial inversion and shift along the coordinate of x is used.

Figure 5 shows the wave profiles observed in the experiment and calculated in the long-wave approximation using the algorithm of accelerated convergence. A good agreement between the experimental (points) and calculated (curves) data can be seen. Note that the results of the numerical-analytical model for two configurations of the base of the vessel (figures 4*a* and 4*c*) completely coincide at b_0 = 0.01 cm. The latter indicates the possibility of obtaining numerical estimates of the displacement of the liquid at the top of the wedge at known wave amplitude on the end wall.

IOP Conf. Series: Materials Science and Engineering 1191 (2021) 012005 doi:10.1088/1757-899X/1191/1/012005

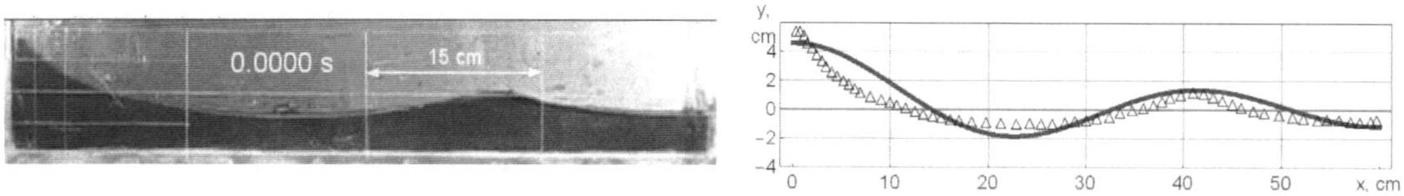

a

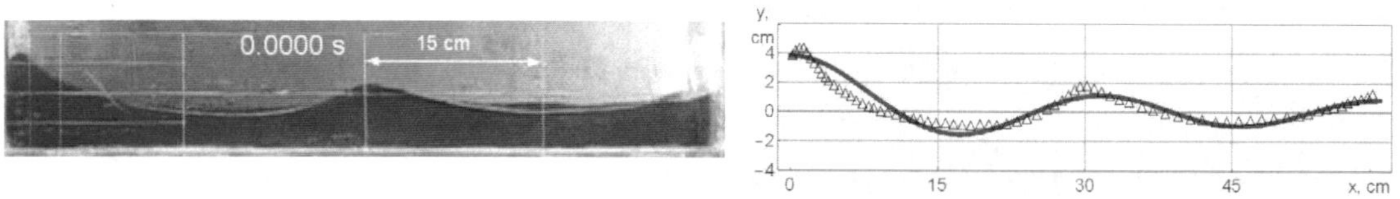

b

Figure 5. Experimental (triangles) and calculated (curves) wave profiles. To solve the boundary value problem (5) the method of accelerated convergence was used: *a, b* – the third and fourth wave modes

The base of a cylindrical vessel in the form of a wedge with a small apex angle α can be considered as a sector of a circle with a central angle α – figure 6*a, b*. According to works [15], in cylindrical polar coordinates (r, θ) the standing wave velocity potential must satisfy

$$\Phi = A\cosh m(z+h)\mathrm{J}_{s\pi/\alpha}(mr)\cos\frac{s\pi}{\alpha}\theta\cos\omega t$$

Here α is the angle of the sector; $\mathrm{J}_{s\pi/\alpha}(mr)$ is the ($s\pi/\alpha$)-th order Bessel function of first kind; the values of m are the positive roots of

$$\left(\mathrm{dJ}_{s\pi/\alpha}(mr)/\mathrm{d}r\right)_{r=l} = 0$$

In our case $s \equiv 0$ – only the radial wave mode is considered, and the standing wave in a vessel shaped like a sector of a circle are determined by

$$\eta\,(r) = A_0\,\mathrm{J}_0\,(m_n r)\cos\omega_n t \qquad (6)$$

where A_0 is the displacement of the liquid free surface at $r = 0$; m_n is any root to the transcendental equation

$$\mathrm{J}_0'\,(m_n l) = 0 \qquad (7)$$

and

$$\omega_n^2 = m_n g\tanh m_n h \qquad (8)$$

We use the parameters of our experiment: $l = 60$ cm and $h = 4$ cm. Then for the third and fourth wave modes $n = 3, 4$ we have the following roots of equation (7)

$$m_{3,4} = 0.170,\ 0.222\ \mathrm{cm}^{-1}$$

The corresponding frequency values are calculated by the formula (8)

$$\omega_{3,4} = 9.910,\ 12.441\ \mathrm{s}^{-1}$$

Figure 6*b* shows the profiles of the third and fourth modes, calculated using formula (6) (curves) and in the long-wave approximation by the method of accelerated convergence (points). It is seen that these two approaches lead to absolutely coinciding profiles of a standing gravity wave in a vessel with a wedge-shaped base.

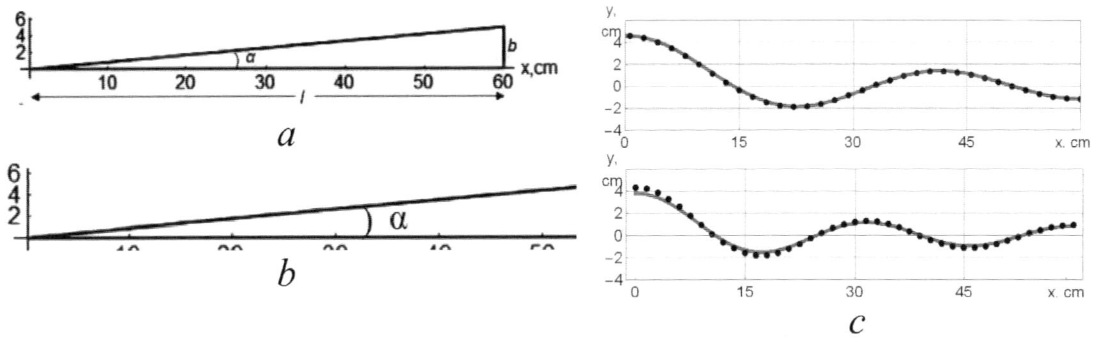

Figure 6. (*a, b*) Wedge and circle sector with the small angle $\alpha = \pi / 36$. (*c*) Wave profiles calculated using formula (6) (curves) and in the long-wave approximation by the method of accelerated convergence (points)

Table 1. Standing wave frequencies.

n	Rectangular	Wedge-shaped	Circular sector
3	8.900	9.953	9.910
4	10.650	11.635	12.441

Finally, table 1 shows the frequency of the standing waves excited in a rectangular vessel, in a vessel with a wedge-shaped base, and in a vessel in the form of a circular sector. For the latter, the frequency was estimated by the formula (8). It can be seen that for the third and fourth modes ($n = 3$, 4), the circular sector model gives frequency estimates close to the experimental values. The frequency of the waves in the wedge-shaped vessel increases.

Note that if the standing wave model for a circular sector requires specifying the displacement of the free surface of the liquid in the sector center, then the accelerated convergence algorithm allows this displacement to be estimated from the wave amplitude on the end wall of the wedge. On the other hand, the model of standing waves in a vessel with a base in the form of a circular sector allows carrying out a kinematic analysis of waves in a linear approximation, as well as, taking into account higher approximations, an estimate of nonlinear effects.

4. Concluding remarks

New experimental data on the excitation of standing surface gravity waves in a cylindrical vessel with a wedge-shaped base are presented. Their fundamental difference from the liquid wave motions in a rectangular vessel is the absence of spatial symmetry for the profile of maximum development and an increase in the wave amplitude towards the wedge tip.

In the long-wave approximation, the problem of standing waves in a vessel of variable width was formulated and numerically solved using the algorithm of accelerated convergence. The calculation results adequately describe the experimental wave profiles for third and fourth modes.

To describe the experimental data, a linear analytical model of standing waves in a cylindrical vessel with a base in the form of a circular sector with a small central angle was successfully used. It is shown that under the conditions of the present experiment, this approach productively complements the numerical estimates of the long-wave approximation.

References

[1] Ibrahim R A 2015 Recent advances in physics of fluid parametric sloshing and related problems *J. Fluids Eng.* **137**(9) 090801 doi:10.1115/1.4029544

[2] Das S P and Hopfinger E J 2008 Parametrically forced gravity waves in a circular cylinder and finite-time singularit *J. Fluid Mech.* **599** 205-228 doi:10.1017/s0022112008000165
[3] Kalinichenko V A 2020 Suppression of intense fluid oscillations by a floating particle layer *Fluid Dynamics* **55**(6) 804–816 doi: 10.1134/S0015462820060063
[4] Kalinichenko V A and Sekerzh-Zen'kovich S Y 2011 Excitation of progressive-standing faraday waves *Doklady Physics* **56**(6) 343–347 doi:10.1134/s1028335811060024
[5] Kana D D 1966 An experimental study of liquid surface oscillations in longitudinally excited compartmented cylindrical and spherical tanks *CR-545 NASA*
[6] Abramson H N, Chu W H and Kana D D 1966 Some studies of nonlinear lateral sloshing in rigid containers *J. Appl.Mech.* **33**(4) 777-784 doi:10.1115/1.36251827
[7] Wilson B 1972 Seiches *Advances in Hydroscience* **8** 1-94 doi:10.1016/b978-0-12-021808-0.50006-1
[8] Friedrichs C T and Aubrey D G 1994 Tidal propagation in strongly convergent channels *J. Geophys. Res.* **99**(C2) 3321–3336 doi:10.1029/93jc03219
[9] Higuera F J, Medina A and Liñán A 2008 Capillary rise of a liquid between two vertical plates making a small angle *Phys. Fluids* **20**(10) *102102.* doi:10.1063/1.3000425
[10] Nesterov S V 1969 Parametric excitation of waves on the surface of a heavy fluid *Morskie Gidrofiz. Issledovaniya* **3(45)** 87–97 (*in Russian*)
[11] Kalinichenko V A, Nesterov S V, Sekerzh-Zen'kovich S Ya and Chaykovskii A A 1995 Experimental study of surface waves with Faraday resonance excitation *Fluid Dynamics* **30(1)** 101–06
[12] Sretenskii L N Theory of wave motions in a fluid. 1977. Moscow. Nauka. 303 p (*in Russian*)
[13] Akulenko L D and Nesterov S V High-precision methods in eigenvalue problems and their applications. 2005. Boca Raton. CRC Press. 255 p
[14] Kalinichenko V A, Nesterov S V and Soe A N 2017 Standing surface waves in a rectangular tank with local wall and bottom irregularities *Fluid Dyn.* **52**(2) 230–238 doi: 10.1134/S0015462817020077
[15] Wehausen J V and Laitone E V 1960 Surface waves *Encyclopedia of Physics/Handbuch Der Physik* **IX** 446–778 doi:10.1007/978-3-642-45944-3_6

Acknowledgments
The work was supported by a government program (Project No. AAAA-A20-120011690131-7)

IOP Conf. Series: Materials Science and Engineering 1191 (2021) 012006 doi:10.1088/1757-899X/1191/1/012006

The Class of Self-Oscillating Systems with Analytically Defined Limiting Cycles

S V Nesterov [1], V G Baydulov [1,2]

[1] Ishlinsky Institute for Problems in Mechanics of the RAS, Moscow, Russia
[2] Bauman Moscow State Technical University, Moscow, Russia

E-mail: bayd@ipmnet.ru

Abstract. It is studied the model that generalizes the well-known equations of the nonlinear oscillation theory (Van der Pol and Rayleigh). The limit cycles of the model are determined by the total energy of the oscillation in the absence of dissipation / energy input into the system. In contrast to the traditional equations of self-oscillations, equations in the Lagrange form are used here. Phase portraits of limit cycles which can be used in engineering are constructed analytically and numerically.

1. Introduction.

The monograph [1] describes a solid-state wave gyroscope, the physical basis of which is a solid-state resonator made of high quality quartz. Undamped standing waves are excited in the resonator, which make it possible to transform this resonator into a gyroscope. A detailed description of the principle of operation is also given in a number of other publications [2–4]. Currently, wave gyroscopes are widely used in technology [5, 6]. The analysis of the properties of such oscillatory systems, the problems of determining their stability and control of parameters are being actively investigated at present [7–10]. The self-oscillating regime of the resonator was considered in detail earlier [11]. There are two thoroughly studied differential equations describing the emergence and existence of steady-state oscillations. These are the Van der Pol equation [12] and the Rayleigh equation [13]. Both of these equations are used to design and analyze the operation of various types of alternators used in electrical and radio engineering. The main mathematical and experimental results are in the instability of the trivial state (rest point) and the existence of a periodic solution with constant amplitude and period. This solution is called a limit cycle.

Consider the hybrid equation obtained from the Van der Pol and Rayleigh equations

$$\ddot{u} - 2\varepsilon\left(1-(1-a)u^2 - a\dot{u}^2\right)\dot{u} + u = 0 \qquad (1.1)$$

Here the parameter a can take on the values $0 \le a \le 1$. For $a = 0$, equation (1.1) goes over into the van der Pol equation, and for $a = 1$, into the Rayleigh equation. If we put $a = 1/2$, then we get the equation proposed by Academician V.F. Zhuravlev

$$\ddot{u} - 2\varepsilon\left(1-\tfrac{1}{2}\left(u^2+\dot{u}^2\right)\right)\dot{u} + u = 0 \qquad (1.2)$$

which has an exact solution for any values of the parameter ε

$$u = \sqrt{2}\cos t, \quad \dot{u} = \sqrt{2}\sin t \tag{1.3}$$

with a period $T = 2\pi$. The limit cycle $u = \sqrt{2}\cos t$ is stable. Note also note that analytical results were obtained for the Van der Pol and Rayleigh equations in the cases when the parameter $\varepsilon \ll 1$

Previously [14], a general equation was constructed that makes it possible to analytically find limit cycles, as well as to study their stability. Здесь будет развито дальнейшее обобщение уравнений Ван дер Поля и Рэлея на случай, когда свободная колебательная система описывается уравнениями Лагранжа.

2. Lagrangian description of the generalized model.

Let u be a generalized coordinate and $L(u,\dot{u})$ a Lagrange function. The equation of free oscillation has the form

$$\frac{d}{dt}\frac{\partial L}{\partial \dot{u}} - \frac{\partial L}{\partial u} = 0 \tag{2.1}$$

Equation (2.1) has the first integral (energy integral)

$$\dot{u}\frac{\partial L}{\partial \dot{u}} - L = E = \text{const}$$

It is assumed that there are exist periodic solutions of equation (2.1).

Let in a mechanical system acts nonlinear friction of the form

$$2\varepsilon\left[E_0 - \left(\dot{u}\frac{\partial L}{\partial \dot{u}} - L\right)\right]\dot{u}, \tag{2.2}$$

which provides instability of the rest point and leads to the appearance of limit cycles. In technology, such a self-oscillation mechanism is implemented in the form of electromechanical devices of various types [15].

The equation of motion of a mechanical system with friction of the form (2.2) is written as

$$\frac{d}{dt}\frac{\partial L}{\partial \dot{u}} - \frac{\partial L}{\partial u} = 2\varepsilon\left[E_0 - \left(\dot{u}\frac{\partial L}{\partial \dot{u}} - L\right)\right]\dot{u} \tag{2.3}$$

Multiplying (2.1) by $\dot{u}$, we obtain the equation for the energy of the system

$$\frac{dE}{dt} = -2\varepsilon(E - E_0)\dot{u}^2 \tag{2.4}$$

where $E = \dot{u}\left(\frac{\partial L}{\partial \dot{u}} - L\right)$.

Equation (2.4) shows that if the initial data are such that $E > E_0$, then the energy turns out to be a non-increasing function of time and vice versa. Thus, the form of the limit cycle is determined by the condition $dE/dt = 0$ ($E = E_0$), which for the Lagrange function is written as

$$\dot{u}\frac{\partial L}{\partial \dot{u}} - L = E_0 \tag{2.5}$$

For an oscillator described by the equation

$$\ddot{u} = -f(u), \; f(-u) = -f(u) \tag{2.6}$$

with energy integral

$$E = \frac{\dot{u}^2}{2} + \Pi, \; \Pi = \int_0^u f(u)du \tag{2.7}$$

accounting nonlinear friction (2.2) leads to oscillation equation

$$\ddot{u} + 2\varepsilon(E - E_0)\dot{u} + f(u) = 0 \tag{2.8}$$

where ε has the meaning of the feedback coefficient, E_0 is a constant that shows how long the energy of the oscillator E can grow so that the negative resistance decreases to zero. If condition (2.5) is considered as a function on the phase plane $(u,\dot{u})$, whose graph forms a closed curve (a set of closed curves – a multiply connected region), then such a region will be the limit cycle of equation (2.8). Let us write the equation of the limit cycle in the explicit form

$$\frac{v^2}{2}+\Pi(u)=\frac{v^2}{2}+\int_0^u f(u)du=E_0\text{, where } v=\dot{u}$$

If we put $v=0$, then the amplitude is determined by the equation

$$\Pi=E_0 \tag{2.9}$$

Here, by default, it is assumed that for a given value of E_0 there exist two finite roots of equation (2.9) u_1 and u_2. Then the oscillation period will be determined by the expression

$$\int_{u_1}^{u_2}\frac{du}{\sqrt{2(E_0-\Pi(u))}}=\frac{T}{2} \tag{1.11}$$

The value of the period and other properties of the oscillations will obviously be determined by the value of the energy parameter E_0 and the form of the function $f(u)$.

3. Examples.

1. An example of a self-oscillating system with a symmetric limit cycle

$$\ddot{u}+2\varepsilon(E-E_0)\dot{u}+u^5\exp(bu^6)=0;\quad E_0=1,\ \varepsilon=1/8,\ b=2$$

$$E=\frac{\dot{u}^2}{2}+\frac{1}{6b}\left(\exp(bu^6)-1\right)$$

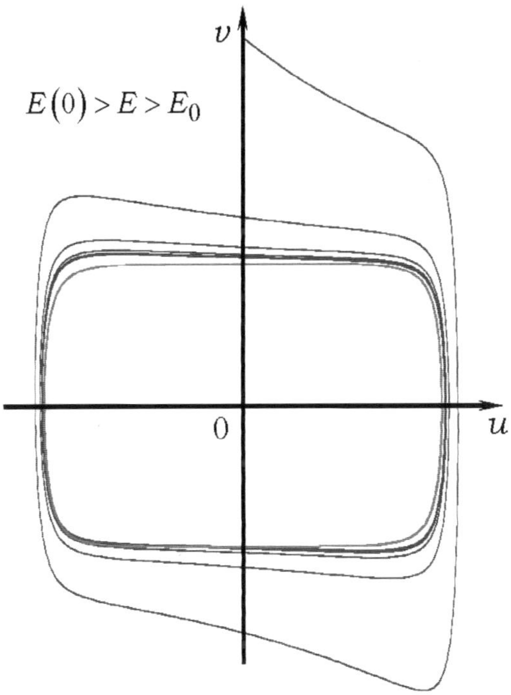

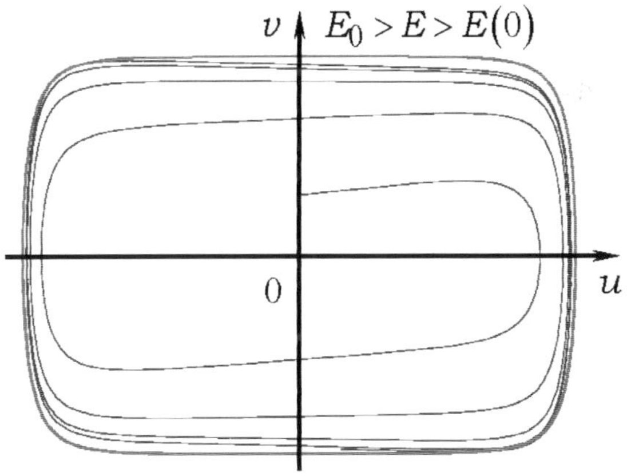

2. An example of a self-oscillating system with an asymmetric limit cycle

$$\ddot{u}+2\varepsilon(E-E_0)\dot{u}+u\exp(bu)=0;\quad E_0=1,\ \varepsilon=1/8,\ b=0.98$$

$$E=\frac{\dot{u}^2}{2}+\left(\frac{u}{b}-\frac{1}{b^2}\right)\exp(bu)+\frac{1}{b^2}$$

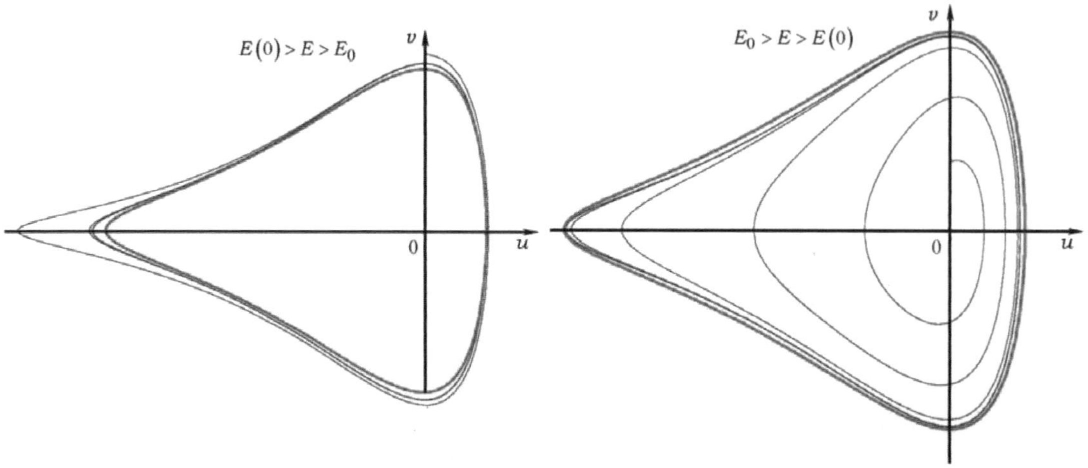

4. References

[1] Klimov D M and Zhuravlev V Ph 1985 *Wave Solid-State Gyroscope* (Moscow: Nauka) p 125 (in Russian)
[2] Klimov D M, Zhuravlev V Ph and Zhbanov Yu K 2017 *Quartz Hemispherical Resonator (Wave Solid-State Gyroscope)* (Moscow: Kim L.A.) p 194 (in Russian)
[3] Zhuravlev V Ph 2011 *Actual Probl. Aviation&Aerospace Syst.: Processes, Models, Experiment,* **2** (33) 118.
[4] Zhuravlev V Ph 2016 *Rus. J. Nonlin. Dyn.* **12**(2) 211.
[5] Yi T, Wu X Z, Xiao D B, Xi X and Tan Y Q 2011 *Appl. Mech.&Mater.* **110–116** 715. DOI: 10.4028/www.scientific.net/amm.110-116.715
[6] Negri C, Labarre E, Lignon C, Brunstein E and Salaün E 2016 *Gyrosc.&Navig.* **7**(3) 223.
[7] Trutnev G A 2019 *Vestn. Udmurt. Univ. Matem. Mekh. Komp'yut. Nauki* **29**(1) 84 (in Russian)
[8] Yi G, Xie Y, Qi Z and Xi B 2015 *Math. Probl. in Engng.* **2015** ID 104041. DOI: 10.1155/2015/104041
[9] Seregin S V 2017 *Rus. J. Nonlin. Dyn.* **13**(3) 423. DOI: 10.20537/nd1703009
[10] Martynenko Yu G, Merkuryev I V and Podalkov V V 2008 *Mech. Sol.* **43** 379.
[11] Zhuravlev V Ph 2020 *Mech. Sol.* **55** 132.
[12] Van der Pol B 1927 *The London, Edinburgh & Dublin Phil. Mag. & J. Sci.* **2**(7) 978.
[13] Strutt J W 1894 *The Theory of Sound*. vol. 1. (London: Macmillan and Co.)
[14] Nesterov S V and Baydulov V G 2020 *Mech. Solids* **55**(8) 1157.
[15] Hensen R H A 2002. *Controlled Mechanical Systems with Friction.* Eindhoven: Univ. Press, https://doi.org/10.6100/IR551394

Acknowledgments
This work was supported by the state assignments AAAA-A20-120011690138-6 and AAAA-A20-120011690132-4.

The effect of perforation on aerodynamic characteristics and the vortex flow field around a flat plate

V T Kalugin[1], A S Epikhin[1,2*], P A Chernukha[1] and M D Kalugina[1]

[1]Bauman Moscow State Technical University, Moscow, Russia
[2]Ivannikov Institute for System Programming of the RAS, Moscow, Russia

[*]E-mail: andrey.epikhin@bk.ru

Abstract. The aim of the present investigation is to obtain and analyse vortex flow field and aerodynamic characteristics of perforated plates at different angles by numerical simulation using OpenFOAM software package. For fully perforated plate's passive gas injection through the holes close to the leading edge induce separation zone deformation and affect the whole flow structure downstream. Perforation along the leading edge tends to result in more monotonous curves of normal force and pitching moment coefficients.

1. Introduction

In aerodynamic applications perforation is used as one of the methods for passive flow control which is primarily implemented on aircraft and missiles. A primary concept of this method is the attenuation of large-scaled vortices generated in the near and far wake of the drag-inducing devices. These vortices, when untreated, interact with other components which results in additional unsteady loading on the components. Purser and Turner [1] investigated the aerodynamic performance of perforated split flaps used on dive bombers. Their results showed that perforated single and double split flaps with porosity σ of 30% provide the required dive maneuverer characteristics and at the same time attenuate buffeting effects. Additional data on the performance of perforated drag brakes was provided in [2]. Again, it was reported that the effect of perforation, "smoothens" the flow behind the flaps and significantly reduces the amplitude of pressure fluctuations. For the level of porosity of 27%, the width of the wake was reduced up to 20%. It was also reported that the drag of the solid flap was 15% higher compared to the drag of the perforated flaps with a round perforation. The perforation of the air brake can be beneficial with regards to reducing unsteady aerodynamic loads [3]. Fundamental work on the aerodynamics of perforated plates located normal to an airstream simulating porous fences was published in [4]. The properties of turbulent flow behind the plates were measured using hot-wire and a pulsed hot-wire apparatus. Later research carried out by Lee and Ko [5] on perforated and solid Gurney and Gurney-type flaps located at the trailing edge of an airfoil demonstrated that perforated flaps with $\sigma > 23\%$ reduce C_D and C_L but have an overall benefit with regards to lift-to-drag ratio and improve pitching moment characteristic of the airfoil. The papers [6,7] confirm integral aerodynamic benefits due to perforation. Detached Eddy Simulation was performed for a high fineness ratio body with deployed perforated and solid drag brakes. Perforated drag brakes were chosen to resolve the problem with the deteriorated performance of the high fineness ratio body with solid brakes. The numerical modelling showed that non-desirable conical motion of the body caused by the interaction of the vortices shedding from the brakes with the after a body was eliminated due to perforation. All the studies report similar benefits of perforation in terms of attenuation of large scale eddies for porosities level of 30% and above. Large-Eddy Simulations of the flow created by an infinite multi-perforated plate was presented in [8].

Paper [9] presents the law for the modelling of the flow through a perforated plate. As well as fractal porosity [10] perforation of the surface can be applied to reducing spoiler noise. Also perforation of deflectors used for further jet noise reduction of a turbofan engine [11]. Other publications include landing gear noise reduction using perforated fairing [12, 13]. So the majority of the authors have also reported overall drag reduction and loss of lift coefficient for perforated drag devices. At the same time, only a limited amount of information is available which helps understand the underlying physics of flow in the near wake of perforated plates to find an optimum configuration.

The objective of the investigation is to assess the aerodynamic performance of solid and perforated inclined plates which simulate forward-opening configuration of the airbrakes. This configuration is considered favourable in terms of reduced interaction between shear layer shedding from the plate and the aircraft components located in the near wake. To achieve this objective, a series of three-dimension calculations were performed and compared with Particle Image Velocimetry (PIV) measurements. A brief review of this study was also reported in [14].

2. Numerical methods

In this paper, the solution of the equations for incompressible gas with the application of the finite volume method is computed using the OpenFOAM software. The PISO algorithm implemented in the pisoFoam solver is used to link the pressure and the velocity field. Large Eddy Simulation (LES) has been used to obtain a detailed representation of the turbulent velocity flow field in the near wake of the plates. This approach directly computes (resolves) energy-containing turbulent scales while small scales are not resolved and should be modelled. To model the interaction of large vortices with fine-scale turbulence, Smagorinsky subgrid model [15] with empirical constant $C_s = 0.15$ has been used. An important feature of the application of eddy-resolving approaches is that the accuracy of vortex flows calculation depends on the differencing schemes of convective flux discretization. In this work, the modified numerical scheme filteredLinearM has been used for convective flux discretization. More details about this scheme can be found in the early works [16, 17]. Crank-Nicolson scheme has been used for time discretization. The time step has been chosen by Courant–Friedrichs–Lewy condition $CFL \approx 0.2$. These selected numerical approaches allow to correctly calculate the process of vortices propagation and dissipation in space and satisfy the requirements for eddy-resolving approaches [17].

3. Computational setup

Three-dimensional drag plates with and without perforation holes were chosen. All the plates have the same dimensions. The span b is 500 mm, the chord c is 150 mm and the thickness is 4 mm. The plates have elliptical leading and trailing edges. The diameter of perforation holes d_j is 4 mm. The holes are drilled uniformly in span wise and chord wise directions, except the region close to the symmetry plane of the plates, this is due to the mounting constraints of the plates in the wind tunnel. Figure 1 shows the representation of the studied plates.

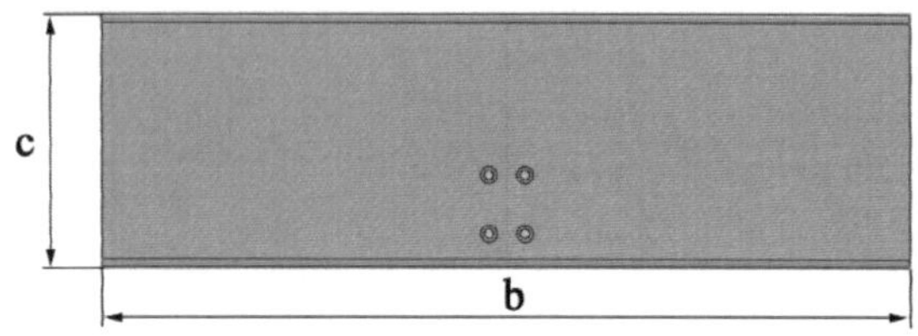

(a)

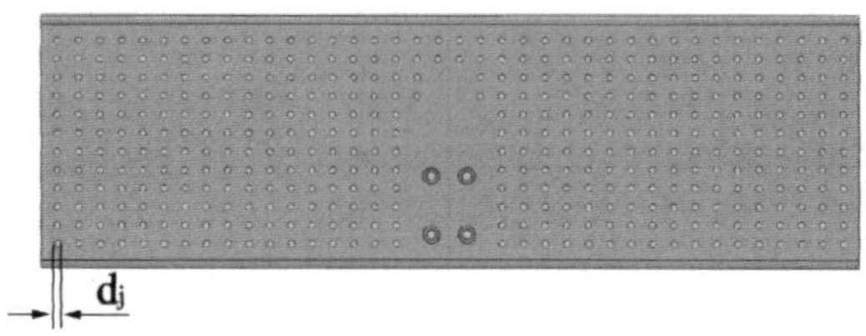

(b)

Figure 1. Schematic representation of the plates: (a) solid; (b) perforated

A configuration of the 3D computational domain built around an inclined plate represents in Figure 2. The distance from the plate midplane to the inlet interface is $25 \cdot c$, the distance from the plate midplane to the outlet interface is $65 \cdot c$. The height and the width of the domain are $50 \cdot c$. An unstructured grid with the total number of cells of $15 \cdot 10^6$ and $30 \cdot 10^6$ was generated for non- and perforated plates respectively using snappyHexMesh mesh generator. To detailed flow structure investigation in the perforation holes, additional semi-3D calculations were performed on the plate with only 3 rows of the perforation holes. This was driven by the limitations of the total number of cells, which were not sufficient for the detailed flow resolution in the holes. The unstructured grid with the total number of cells of $21 \cdot 10^6$ was generated around the plate with the semi-3D perforation. The representation of the mesh fragment around a perforation hole is given in Figure 3.

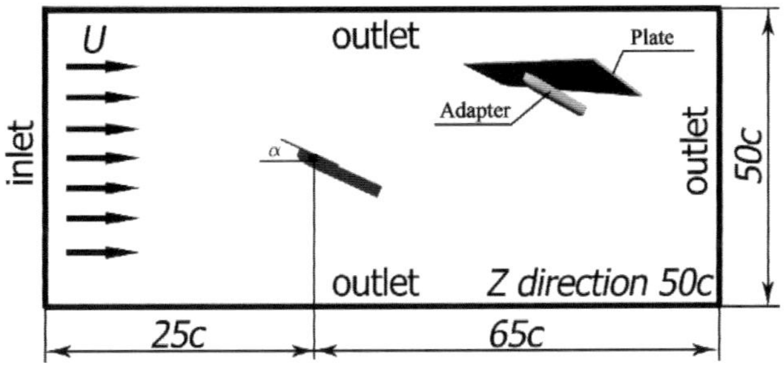

Figure 2. The computational domain

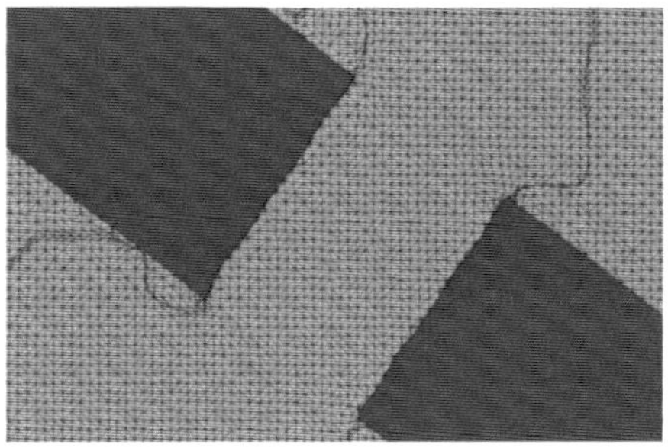

Figure 3. The mesh around the perforation hole

At the surfaces of the perforated and solid plate were applied no-slip boundary condition. A damping function suggested by van Driest [18] has been employed near the wall. At the inlet boundary plane, the velocity vector field gradually increased from zero to the free stream velocity U_∞ equal to 30 m/s. This allows getting rid of solution oscillations and increases numerical stability during the initial phase of the simulation. For the semi-3D simulation periodicity in the spanwise direction was applied for the left and right boundaries of the fluid domain. The calculations were performed for the free stream velocity corresponding to Re_c of $3.0 \cdot 10^5$.

4. Results and Discussion

4.1. Vortex flow field calculation

Figures 4 and 5 show the flow structures of the detached flow near the solid plate streamlined at angles of attack $\alpha = 18°$ and $24°$.

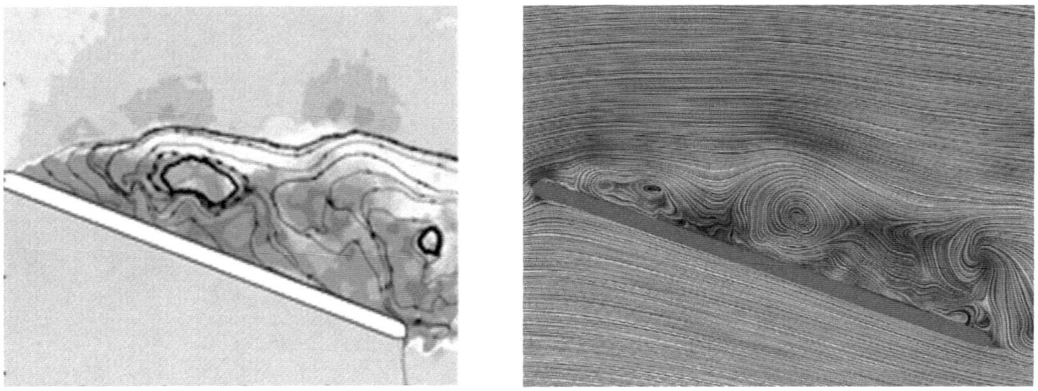

(a) (b)

Figure 4. Instantaneous velocity flow field for a solid plate at $\alpha = 18°$: (a) PIV experiment; (b) numerical simulation

IOP Conf. Series: Materials Science and Engineering 1191 (2021) 012007 doi:10.1088/1757-899X/1191/1/012007

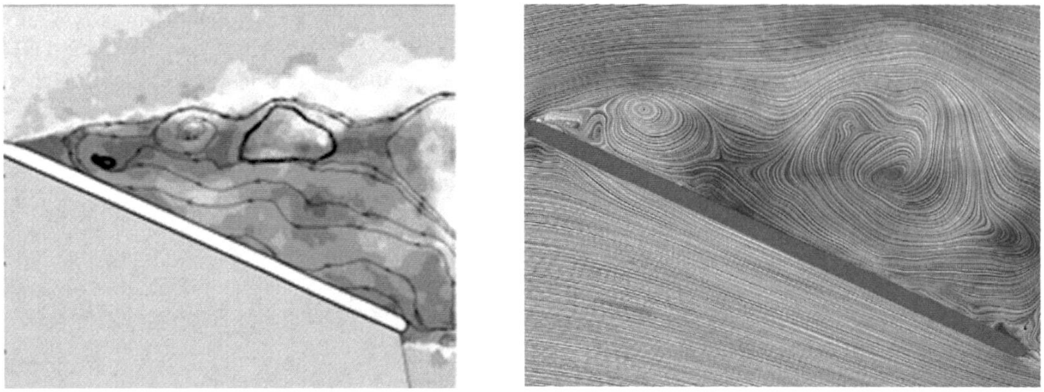

(a) (b)

Figure 5. Instantaneous velocity flow field for a solid plate at α = 24°: (a) PIV experiment; (b) numerical simulation

At angle of incidence of 18°, the shear layer developing from the leading edge is unstable and propagates towards the surface of the plate thus destructing fluid circulation in the clockwise direction. A sequence of small and mid-scale vortices form along the chord of the drag plate. The flow structures at angle of incidence of 24° show lower vorticity level at the surface of the plate and slower decaying of the shear layer shedding from the leading edge. As it can be seen at the angle of incidence of 24°, the shear layer forming at the leading edge of the plate rolls up in the sequences of the vortices shedding downstream. It worth noticing that the calculated vorticities contours at the studied angles of incidence showed that the flow fields at the distance of *b*/4 from the mid plane of the plates are not affected by the tip vortices.

LES simulation of instantaneous flow structures on the leeward side of the plate in comparison with the PIV flow field as well as the flow structure inside the perforation holes presented in Figure 6. As shown in Figure 7, the intensity of the jets reduces with increased distance from the leading edge due to decreased pressure difference between the leeward and the windward sides of the plate.

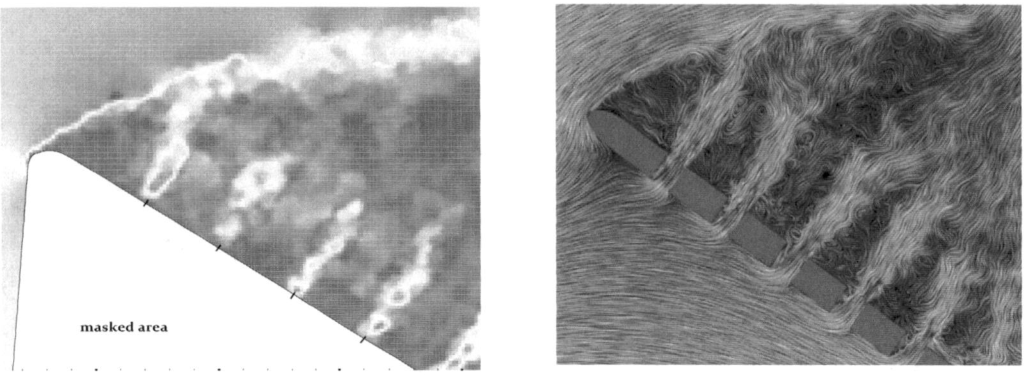

(a) (b)

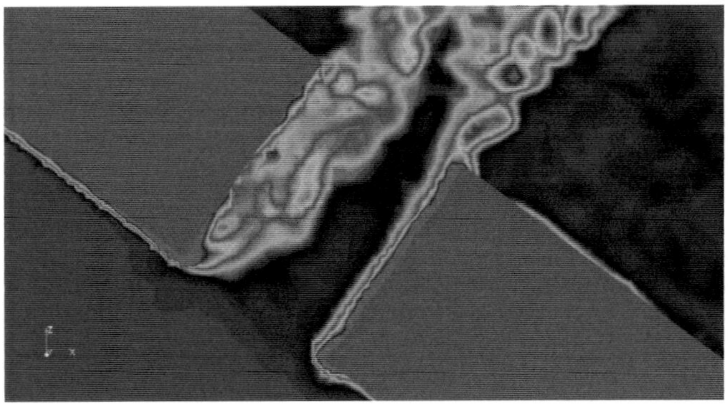

(c)

Figure 6. Instantaneous flow fields for perforated plate at α = 30°: (a) PIV experiment; (b) numerical simulation; (c) vorticities within the hole

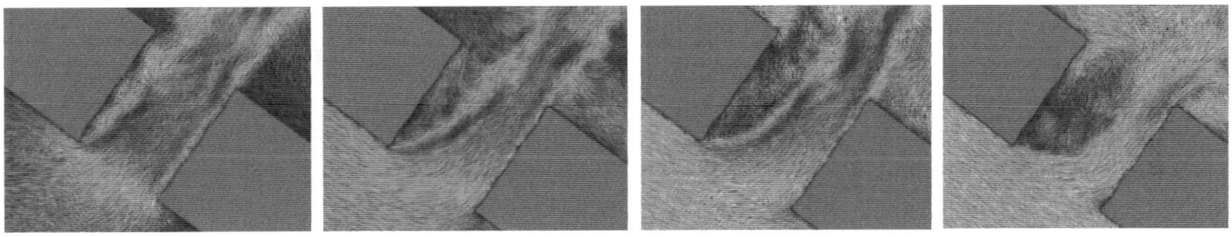

(a) (b) (c) (d)

Figure 7. Instantaneous flow structure within the holes, depending on the distance from the leading edge

The flow separates from the sharp edges at the holes' inlets and reduces the effective area of the holes. This effect was simulated and reported in paper [19]. The velocity at the inlets of the perforation holes located close to the leading edge, i.e. close to the area of flow stagnation, is nearly aligned with the axis of the holes (Fig. 7a). Therefore, the effective area of the hole is larger. Further, towards the trailing edge of the drag plate, the direction of the flow is parallel to the surface of the plate and the flow needs to turn more at the holes' inlets (Fig. 7b,c). This causes larger flow separation and reduction of the effective area. Close to the trailing edge, the intensity of the jets is the weakest as the flow separation inside the hole is the largest (Fig. 7d). A local area of the flow stagnation at the hole inlets located close to the trailing edge can be seen in the bottom right (Fig. 7c,d). For a larger diameter of the perforation hole at the same angle of incidence the effective area of the holes is larger, therefore the maximum achievable jet velocity will be higher. With a change in the plate thickness, the degree of filaments deformation at the outlet from the holes increases, they cease to be co-directional with the main flow. This is due to the increase in the surface area of the inner wall of the holes. An inhomogeneous flow structure is observed within the holes, and local separation zones are formed on the lateral sides.

4.2. Aerodynamic performances

Figure 8 summarizes changes in drag and lift coefficients due to perforation. The absolute values of drag and lift coefficients are plotted against the angle of incidence α. The presented results show that the C_D values with perforation are higher than the C_D for the solid plate. The increase of the drag coefficient for perforated plates can be explained by surface discontinuities which contribute to the total projected area and complex flow development inside the holes. Within all studied α range C_L values of the solid plate are higher than C_L values of the perforated plates. The $C_m(\alpha)$ curves for perforated plates increase linearly with increased angle of incidence overall studied range of α. For the solid plate, at the angles of incidence up to 12°, the pitching moment coefficient has a higher gradient followed by a sudden drop in at the angles of incidence corresponding to full separation from the leeward side of the plate. In contrast to a solid one, the curves of aerodynamic performances are more monotonic for a perforated plate due to the ongoing transformation of the flow and the formation of small local separation zones.

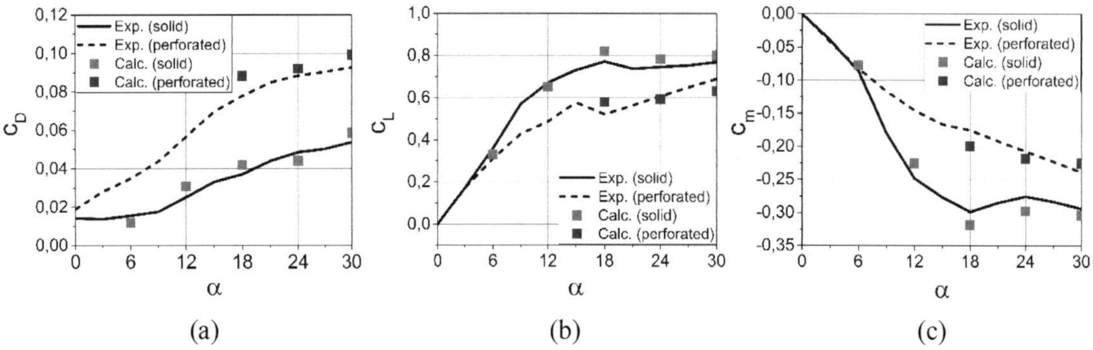

Figure 8. Aerodynamic performances as a function of angle of attack: (a) drag; (b) lift; (c) pitch moment. Lines show experimental data (dash is a perforated plate). The symbols represent calculation results (red is solid plate and blue is perforated)

5. Conclusion

Flow fields in the near wake of solid and perforated inclined drag plates were investigated by numerical simulation and compared with PIV and balance measurements. The calculation aerodynamic characteristics of the solid plate are in good agreement with experimental values. The increase in drag coefficient for perforated plates at high angles of incidence is 60%-70%. At the same time, loss of lift coefficient in comparison with the solid plate is observed. The difference between C_L values is the maximum at the angle of incidence of 18° and reduces with increased α, e.g. at the angle of incidence of 30° the ΔC_L values are 11%.

The flow field on the leeward side can be split into two distinct areas with perforated plates. The area close to the leading edge is dominated by the jets mixing with the shear layer developing from the leading edge. For the angle of incidence of 18° the jets mixing with the shear layer reduces instability of the separation flow. The area close to the trailing edge is dominated by a clockwise rotating vortex which is shifted downstream from the surface of the plates. The intensity of the passive jets is determined by the pressure difference between the leeward and windward sides of the plate and the angle between the flow approaching the holes' inlets and the axis of the hole. The strongest jets were detected to develop at the leading edge while the weakest ones are formed by the air passing through the holes at the trailing edge.

Acknowledgments

All calculations were carried out using the equipment of the shared research facilities of HPC computing resources at Lomonosov Moscow State University. The authors acknowledge the Deutscher Akademischer Austausch Dienst (DAAD) exchange programme for the provided fellowship funding and DLR-Gottingen for the permission to publish the data.

Reference

[1] Purser P E and Turner T R 1941 *Langley Memorial Aeronautical Laboratory* (VA, United States) NACA report WR-L-445.
[2] Davies H and Kirck F 1951 *Aeronautical Research Council Reports and Memoranda* (London) A.R.C Technical Report 2614(5979).
[3] Epikhin A 2021 *Progress in Computational Fluid Dynamics* **21(2)**.
[4] Castro I P 1971 *Journal of Fluid Mechanics* **46(3)** 599-609.
[5] Lee T and Ko L 2009 *Experiments in Fluids* **46(6)** 1005–19.
[6] Cummings R, Divine J, Yechout T, Wetlesen D and Kealos J 2006 *44th AIAA Aerospace Sciences Meeting and Exhibit (Nevada)* AIAA Paper 2006-668.

[7] Florendo C, Yechout T, Siegel S, Cummings R and Kealos J 2007 *Journal of Spacecraft and Rockets* **44(3)** 589-596.
[8] Mendez S, Nicoud F and Poinsot T 2007 *Lecture Notes in Computer Science* **56** 289-303.
[9] Doerffer P and Bohning R 2000 *Aerospace Science and Technology* **4(8)** 525-534.
[10] Nedić J, Ganapathisubramani B, Vassilicos J, Borée J, Brizzi L and Spohn A 2012 *AIAA Journal* **50(12)** 2695-2710.
[11] Papamoschou D 2009 *AIAA Journal* **47(6)** 1408-20.
[12] Boorsma K, Molin N, Zhang X and Chow L 2009 *AIAA Journal* **47(1)** 33-43
[13] Boorsma K, Zhang X, Molin N (2008) *14th AIAA/CEAS Aeroacoustics Conference (Canada)* AIAA Paper 2008-2961.
[14] Golubev A, Stolyarova E and Kalugina M 2019 *AIP Conference Proceedings* **2171** 130002.
[15] Smagorinsky J 1963 *Monthly Weather Review* **91(3)** 99-165.
[16] Epikhin A and Kalugin V 2018 *IOP Conf. Ser.:Mater. Sci. Eng.* **468** 012035.
[17] Epikhin A 2019 *Mathematical Models and Computer Simulations* **11(6)** 1019-31.
[18] van Driest E 1956 *Journal of the Aeronautical Sciences* **23** 1007-11.
[19] Xiong J, Johnson A, Liu F and Papamoschou D 2012 *AIAA Journal* **50(11)** 2525-35.

IOP Conf. Series: Materials Science and Engineering 1191 (2021) 012008 doi:10.1088/1757-899X/1191/1/012008

Analytical estimates of elastic vibrations of separated flaps construction of fairings' transformable assembly-protection unit

V N Bakulin[1], S V Borzykh[2], K V Tereshina[2] and Yu N Shchiblev[2]

[1] Institute of Applied Mechanics of the Russian Academy of Sciences, Leningradsky avenue 7, Moscow, 125040, Russia
[2] S.P. Korolev Rocket and Space Corporation Energia, 4A Lenin Street, Korolev, Moscow area, 141070, Russia

E-mail: vbak@yandex.ru

Abstract. The separation of the fairings protective flaps of the assembly-protection unit is one of the key operations in the zero-lift trajectory. In this regard, the development of the flaps separating process models, at most concerning the process features and the properties of the flaps, in particular, their elastic properties, becomes urgently important. Kinematic schemes of the flaps separation assume their rotation relative to the spin axes fixed on the launcher, and the separation means forces create the moment necessary for the turn, and also cause the flaps vibrations. Small oscillations can be represented as series expansion (decomposition in a row) by eigenforms. A number of analytical estimates of the flaps vibrations level in their separation process are obtained.

1. Introduction

The dynamics modeling of modern space structures was considered, for example, in works [1-3]. The mechanical transformation processes of rocket and space systems, accompanied by their structure and/or configuration changes are critically important operations. The failure or abnormal functioning of them almost always means the failure of the mission tasks [4-7]. The separating batteries processes of rocket-space systems is a typical example of such processes, that's why there are high demands for ensuring security of the separation processes and the validity of choosing characteristics for the separation processes means (CO). At the same time, during ground-based experimental testing, it is not possible to reproduce real flight conditions. Therefore, the main mean for confirming the correctness of adopted technical decisions is mathematical modeling with maximum consideration of the characteristic features of the separation processes [8-12].

2. Statement of the problem

One of the most important types of separation processes is the process of separating the fairings protective flaps of the payload. The fairing is aimed to protect the payload from wind, dust on the launch pad, from intense loads (acoustic, thermal, etc.) at the zero-lift trajectory. The protective function of fairing loses its necessity after passing through the atmosphere, and it separates from the launch vehicle. The fairing has a considerable length, so it is constructed as a composite structure and includes several separately dropped flaps (Fig. 1). In the first phase of the process the flaps are rotate around the axis placed on the vehicle to ensure the organized movement of the separation. Then, after reaching a certain angle of rotation, the connection with the vehicle is broken, and the section of flap autonomous movement begins.

Usually the fairing consists of a number of rotation shells - cylinders, cones, spheres. Under it, with a certain clearance, there is an area of the spacecraft structure. The size of the clearance is determined mainly by the amplitude of the fairing flaps elastic vibrations during separation process. Insufficient clearance may cause the flaps to collide with the machine structure and cause damage. An excessively large clearance ("we carry air") increases the diameter of the fairing, what leads to an increase in aerodynamic drag, loads on the structure, and weight.

The flaps have large weights and moments of inertia about the rotation axes. Flaps turn happens under the influence of significant concentrated forces created by special separation means placed in several points of the structure (Fig. 1). Existing approaches to calculating the dynamics of fairing separation were usually based on flaps models as absolutely solid bodies [8]. In this setting, the motion of the separated flaps was determined relative to the launcher to make sure that there were no collisions with it in the further relative motion. Flaps speed after the separation means were activated and their relative motion trajectories were well consistent with the solid model. But sufficiency of the required clearance, determined by the level of vibrations, was confirmed on ground-based experimental installations for testing the separation process. For choosing optimal clearance we need a predictive calculation of the flaps elastic vibrations level during separation while preliminary design stage.

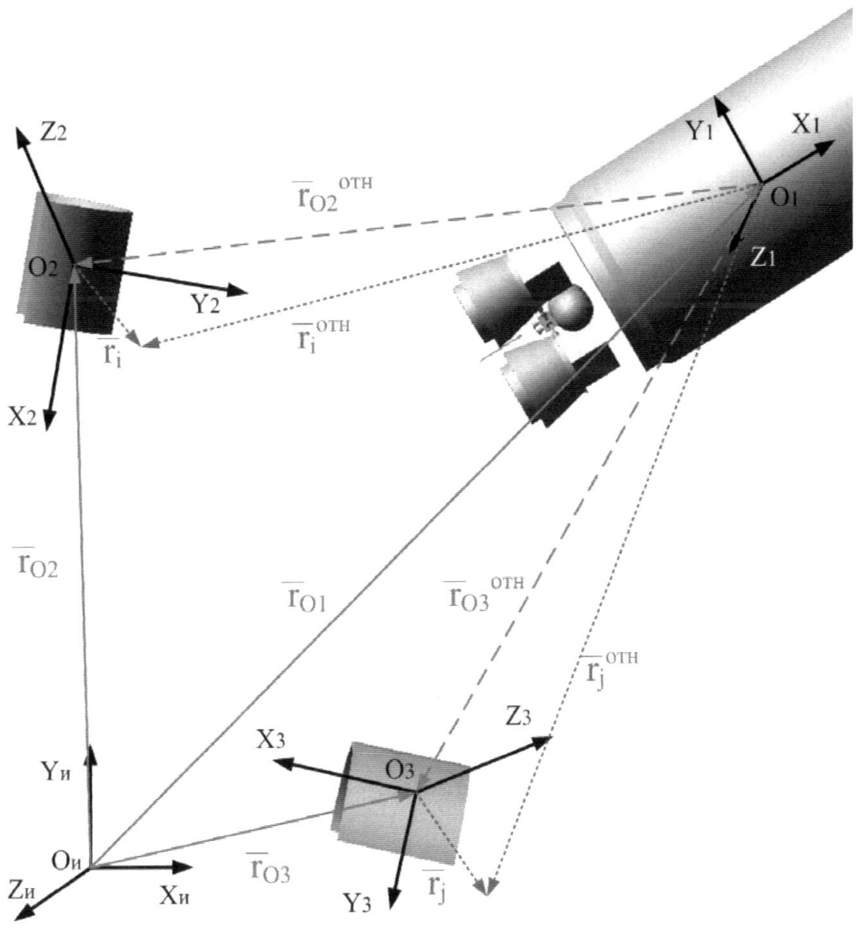

Figure 1. Relative movement of the flaps and the launch vehicle.

3. Materials and methods of research

In the most general formulation, the study of the separation and directional movement of elastic structures is a difficult task, as such movements are described by bulky "hybrid" systems that include both partial differential equations and ordinary differential equations [13]. Based on the structural features of the separated bodies and the experience of the technical implementation of the separation processes, it is possible to make an assumption about the smallness of elastic deformations in comparison with typical linear dimensions of the flaps, reaching tens of meters. Thus, the problem of elastic deformations can be solved in a linear formulation. Therefore it seems rational to approach using the elastic body motion decomposition into natural components: flap movement as a rigid body (described in General case as nonlinear equations) and small linear elastic vibrations near generating motion, which represented in modal form, i.e. as decomposition into natural vibration modes, the determining methods of which are described in the literature [14]. The impact of every particular mode on the common deformation decreases rapidly with increasing mode number because the frequency spectrum of the flaps is usually sparse. So, we can use a finite number of lower modes in practical calculations.

As was noted above, the "solid" models allow us to calculate the relative motion of launcher and flaps after their separation precisely enough. It reflects the fact that inertia forces caused by the flaps movement as a whole practically have no impact on their vibrations. In this regard, the equations of motion as a whole (for solid forms) and small elastic vibrations can be analyzed independently.

The security analysis of flaps separation is based on relative motion trajectories and velocities study as a solid whole and a launcher (Fig. 1)

For each body of the system, the laws of change in the amount of motion and kinetic moment are recorded

$$m_i \frac{d^2 \vec{r}_{o_i}}{dt^2} = \Sigma \vec{F}_i^{(e)} \tag{1}$$

$$\left[I_{o_i}\right]\frac{\vec{\omega}_i}{dt} + \vec{\omega}_i \times \left(\left[I_{o_i}\right]\vec{\omega}_i\right) = \Sigma \vec{M}_{o_i}\left(\vec{F}^{(e)}\right), \tag{2}$$

where m_i, $\left[I_{o_i}\right]$, $\vec{\omega}_i$, $\Sigma \vec{F}_i^{(e)}$, $\Sigma \vec{M}_{o_i}\left(\vec{F}^{(e)}\right)$ - masses, inertia tensors, angular velocities, main vectors and main moments of the external forces of the launch vehicle and the fairing flaps.

Below are the analytical kinematic relations that allow us to determine the radii-vectors of velocity and acceleration of the flaps centers of mass in the coordinate system of launch vehicle.

$$\vec{r}_{o_2}^{\,rel.} = \left[M_{in.\to 1}\right]\left(\vec{r}_{o_2} - \vec{r}_{o_1}\right)$$

$$\vec{V}_{o_2}^{\,rel.} = \left[M_{in.\to 1}\right]\left(\frac{d\vec{r}_{o_2}}{dt} - \frac{d\vec{r}_{o_1}}{dt}\right) - \vec{\omega}_1 \times \left[M_{in.\to 1}\right]\left(\vec{r}_{o_2} - \vec{r}_{o_1}\right) \tag{3}$$

$$\vec{a}^{rel.}_{o_2} = \left[M_{in.\to 1}\right]\left(\frac{d^2\vec{r}_{o_2}}{dt^2} - \frac{d^2\vec{r}_{o_1}}{dt^2}\right) - \vec{\varepsilon}_1 \times \left[M_{in.\to 1}\right]\left(\vec{r}_{o_2} - \vec{r}_{o_1}\right) -$$
$$- \vec{\omega}_1 \times \left\{\vec{V}^{rel.}_{o2}\left[M_{in.\to 1}\right]\left(\frac{d\vec{r}_{o_2}}{dt} - \frac{d\vec{r}_{o_1}}{dt}\right)\right\}$$

Similar parameters of some characteristic "dangerous" flaps construction point (in the sense of possible collision with a launcher) are determined by the dependencies

$$\vec{r}^{rel.}_{i} = \left[M_{in.\to 1}\right]\left(\vec{r}_{o_2} - \vec{r}_{o_1}\right) + \left[M_{2\to 1}\right]\vec{r}_i$$
$$\vec{V}^{rel.}_{i} = \left[M_{in.\to 1}\right]\left(\frac{d\vec{r}_{o_2}}{dt} - \frac{d\vec{r}_{o_1}}{dt}\right) + \left[M_{2\to 1}\right]\left(\vec{\omega}_2 \times \vec{r}_i\right) -$$
$$- \vec{\omega}_1 \times \left\{\left[M_{in.\to 1}\right]\left(\vec{r}_{o_2} - \vec{r}_{o_1}\right) + \left[M_{2\to 1}\right]\vec{r}_i\right\} \tag{4}$$
$$\vec{a}^{rel.}_{i} = \left[M_{in.\to 1}\right]\left(\frac{d^2\vec{r}_{o_2}}{dt^2} - \frac{d^2\vec{r}_{o_1}}{dt^2}\right) + \left[M_{2\to 1}\right]\left\{\vec{\varepsilon}_2 \times \vec{r}_i + \vec{\omega}_2 \times \left(\vec{\omega}_2 \times \vec{r}_i\right)\right\} -$$
$$- \vec{\varepsilon}_1 \times \vec{r}^{rel.}_i - \vec{\omega}_1 \times \left(\vec{\omega}_1 \times \vec{r}^{rel.}_i\right) - 2\vec{\omega}_1 \times \vec{V}^{rel.}_i$$

The transition matrices between coordinate systems included in the above relations are found by traditional methods [8, 13].

The flap motion relative to the launcher can be immediately described in its mobile coordinate system:

$$m_2 \frac{d^2\vec{r}^{rel.}_{o_2}}{dt^2} = \sum \vec{F}^{(e)}_2 + \vec{\Phi}_{2_e} + \vec{\Phi}_{2cor.}, \tag{5}$$

where the inertia forces of the portable motion and the inertia forces of the Coriolis are defined as follows

$$\vec{\Phi}_{2_e} = -m_2\left[\vec{\omega}_2 \times \left(\vec{\omega}_2 \times \vec{r}^{rel.}_{o_2}\right) + \vec{\varepsilon}_2 \times \vec{r}^{rel.}_{o_2}\right]$$
$$\vec{\Phi}_{2cor.} = -m_2\vec{a}_{2_{cor.}} = -2m_2\vec{\omega}_2 \times \vec{V}^{rel.}_{0_2}$$

Vectors $\vec{r}^{rel.}_{o_2}$ and $\vec{V}^{rel.}_{0_2}$ are determined by the expressions (3).

The process of separating the flaps is considered safe in the absence of collisions of any point of the design of the flaps with the launcher in their relative motion.

Intense vibrations of the flaps occur under the action of forces created by special means of separation-pushers (spring, pyrotechnic, pneumatic), solid-fuel engines. Small elastic vibrations can be represented in a modal mode, i.e. in the mode of a series expansion according to the proper modes of vibrations [14,15]. It is assumed that the orthogonality property of the modes holds, and the contribution of each particular mode to the total deformation does not depend on the contribution over the other modes.

Then separate equations for each vibrations tone (of the elementary oscillator)

$$\ddot{S}_n + \omega_n^2 S_n = \sum_i \left[F_i^x f_{ni}^x + F_i^y f_{ni}^y + F_i^z f_{ni}^z \right]$$

will be independent, and in this case they are solved analytically:

$$S_n(t) = S_{n0} \cos \omega_n t + \frac{\dot{S}_{n0}}{\omega_n} \sin \omega_n t + \\ + \frac{1}{\omega_n} \int_0^t \sum_i \left[F_i^x f_{ni}^x + F_i^y f_{ni}^y + F_i^z f_{ni}^z \right] \sin \omega_n (t-\tau) d\tau, \tag{6}$$

where S_{n0}, $\dot{S}_{n0}$ - the initial conditions, F_i- force components i department facilities, modes f_{ni} orthonormal.

The forces in the right-hand sides of equation (7) are usually generated by either pushrods or solid-fuel rocket engines (SFRE). The dependence of SFRE thrust on time is close to constant. The same can be said about the change in the force of pushers, especially in the initial, most dangerous phase of separation process. Then in (7) the forces projections can be taken out of the sign of integral.

From the point of view of reducing perturbations during separation and reducing loads on communication nodes, symmetric process schemes are preferred, when the separation plane coincides with the plane of separated structure symmetry, while the separation means are installed in pairs symmetrically [8, 10]. The separation of the fairing flaps is made exactly according to this scheme.

Let it be $F_1^x = F_2^x;\ F_1^y = -F_2^y;\ F_1^z = F_2^z$ (the x-axis is longitudinal, the z-axis is in the separation plane, the y-axis complements the system to the right). Then $f_{n1}^x = f_{n2}^x;\ f_{n1}^y = -f_{n2}^y;\ f_{n1}^z = f_{n2}^z$ for symmetric modes and $f_{n1}^x = -f_{n2}^x;\ f_{n1}^y = f_{n2}^y;\ f_{n1}^z = -f_{n2}^z$ for skew-symmetric.

SFRE may have a time delay in switching on relative to each other. A spreading of forces developed by individual pushers is also possible. It can be seen from (7) that skew-symmetric forms will be excited only if there is a delay in Δt in switching on of one of the SFRE (or in the difference in forces of pushers), otherwise the excitation covers only symmetric modes. In this case, if Δt=0 for symmetric forms in the period of the force action, we obtain from (7):

$$S_n(t) = \frac{2\left[F_1^x U_{n1}^x + F_1^y U_{n1}^y + F_1^z U_{n1}^z \right]}{\omega_n^2} \psi_n(t); \\ \psi_n(t) = \begin{cases} 0,\ t \le 0 \\ 1 - \cos \omega_n t,\ 0 < t \le T \\ \cos \omega_n (t-T) - \cos \omega_n t,\ t \ge T, \end{cases} \tag{7}$$

where T - SFRE working hours. During the operation of the SFRE, fluctuations occur near the offset position:

$$S_n = \frac{2\left[F_1^x f_{n1}^x + F_1^y f_{n1}^y + F_1^z f_{n1}^z \right]}{\omega_n^2} \tag{8}$$

and after the end of the operation of the separation means, residual fluctuations around the zero value remain S_n=0.

Elastic deformation at an arbitrary point of the flap

$$\begin{Bmatrix} u_i^x \\ u_i^y \\ u_i^z \end{Bmatrix} = \sum_{n=1}^{n_k} S_n(t) \begin{Bmatrix} U_{ni}^x \\ U_{ni}^y \\ U_{ni}^z \end{Bmatrix}, \tag{9}$$

where U_{in}^x, U_{in}^y, U_{in}^z – components of the eigenform vector at this point, summation is performed only on symmetric modes.

Since the contribution to the elastic deformation of higher forms decreases rapidly with the growth of the shape number, a limited number of them is taken into account in real calculations. The maximum value of the elastic strain is obtained by doubling (9).

4. Conclusion

The introduction of simplifying assumptions made possible to consider individual oscillation equations as independent and to obtain analytical estimates (7) - (9). It was assumed, in particular, that flaps vibrations are small, linear, and their angular velocity during their turn is at least an order of magnitude lower than the natural vibration frequencies even in the lowest modes. This made it possible to take into account only the significant magnitude forces, created by separation means, to contribute to the elastic deformation. In the general case, the equations of vibrations for each tone can be related to each other and to the equations of motion of solid bodies by means of expressions of the forces entering their right-hand sides. As a rule, in this case, the elastic bodies separation is modeled using problem-oriented software packages, for example, as MSC Adams or Euler.

Acknowledgments: This work was performed as part of a state assignment, state registration number AAAA-A19-119012290177-0

References

[1] Bakulin V N, Borzykh S V, Voronin V V 2011 The mathematical simulation of the spacecraft landing in the area of its contact with surface Mathematical modeling of the process of landing a spacecraft at the site of its contact with the surface *Aerospace MAI Journal* **18**(4) 38

[2] Bakulin V N, Kokushkin V V, Borzykh S V, Voronin V V 2012 Dynamics of process of landing of the spacecraft winh an landing gear consisting of arm and cord *Aerospace MAI Journal* **19**(5) 45

[3] Bakulin V N, Borzykh S V, Voronin V V 2016 Space vehicle landing dynamics at failure of landing gear *Russian Aeronautics* **59**(1) 23 DOI: 10.3103/S1068799816010049.

[4] Aslanov V, Kruglov G, Yudintsev V 2011 Newton–Euler equations of multibody systems with changing structures for space applications *Acta Astronautica* **68**(11-12) 2080 doi:10.1016/j.actaastro. 2010.11.013

[5] Bakulin D V, Borzykh S V, Ososov N S, Shchiblev Yu N 2004 Modeling the process of opening large-sized solar batteries *Mathematical modeling* **16**(6) 88

[6] Yudintsev V V 2012 Modeling of the processes of disclosure of multi-element structures of spacecraft *Flight* (5) 28

[7] Bakulin V N, Borzykh S V, Ilyasova I R 2011 Mathematical simulation of opening process of multy-element solar battery *Aerospace MAI Journal* **18**(3) 295

[8] Kolesnikov K S, Kozlov V I, Kokushkin V V 1977 *Separation of aircraft stages* (Moscow: Mashinostroenie) p 223

[9] Bogomolov N V, Anfalov A S, Borzykh S V and Bakulin V N 2019 Simulation of process of small satellites separation from deployer installed on cargo spacecraft *Journal of Physics: Conference Series* **1392** 012003 doi:10.1088/1742-6596/1392/1/012003
[10] Bakulin V N, Borzykh S V, Reshetnikov M N 2011 Modeling the relative motion of the reentry capsule and the transport ship during their separation *Aerospace MAI Journal* **3** 287- 94
[11] Bakulin V N, Bogomolov N V, Borzykh S V 2019 Separation Algorithm of the Soyuz MS Spacecraft from Nonstabilized International Space Station *Russian Aeronautics* **62**(4) 577
[12] Bakulin V N, Borzykh S V 2020 Modeling the dynamics of the process of opening large-sized transformable space structures *Russian Aeronautics* **62**(4) 577
[13] Dokuchaev L V 1987 *Nonlinear dynamics of aircraft with deformable elements* (Moscow: Mashinostroenie) p 232
[14] Biderman V L 1980 *Applied theory of mechanical vibrations* (Moscow: Mashinostroenie) p 405
[15] Ilyin M M, Kolesnikov K S, Saratov Yu S 2003 *Theory of oscillations* (Moscow: Publishing House of the Bauman Moscow State Technical University) p 271

IOP Conf. Series: Materials Science and Engineering 1191 (2021) 012009 doi:10.1088/1757-899X/1191/1/012009

Simulation Analysis of Structural Thermoelastic Stresses in Fibrous Composite Materials

S L Kosachev

Bauman Moscow State Technical University, 105005, Moscow, Russia

E-mail: s_kosachev@bmstu.ru

Abstract. This paper presents an analytical method for simulation analysis of structural thermoelastic stresses acting within a uniaxially reinforced fibrous composite material during its manufacture. The method uses a model of a basic fibre-reinforced composite having its geometry and stress state being fully determined by the primitive cell microstructure. For analytical calculation of residual stresses, an experimentally-derived thermoelastic analogy is used based on which residual stresses are considered as proportional to the difference in the thermal expansion coefficients of the matrix and the reinforcing fibres and the temperature difference during cooling of the composite. Using the proposed method, the results of calculations for fiberglass having a hexagonal lattice are presented, and the optimal degree of fibre-reinforcement at which the residual stresses are minimal for various curing temperatures is determined.

1. Introduction

Fibrous composite materials (FCM) are one of the most promising materials used in the fabrication of modern engineering structures. However, structural analysis of FCM structures presents multiple problems, one of which is the occurrence of residual stresses in the material structure.

An FCM structure is usually formed under conditions of elevated temperatures. Subsequent cooling provides the maximum mechanical properties. The cooling process is accompanied by thermal deformations of the FCM components. Since the fibres and the matrix have different thermoelastic properties, they undergo constrained deformation proportionally to the difference in their thermal expansion coefficients and the cooling mode temperature range. Thermoelastic residual stresses are such that at high degrees of fibre-reinforcement and cooling temperature range, the polymer matrix may break even at the structure fabrication stage without applying an external load.

To study the residual stress field in an FCM, various experimental methods (such as polarisation-optical) and various analytical methods are usually used [1-5]. This paper presents an analytical method for calculating structural residual stresses using a basic fibre-reinforced FCM model.

2. Problem statement

The FCM model used in this study is a three-dimensional isotropic piecewise homogeneous medium having elastic and geometric properties that are invariable in the x_3 direction that is doubly periodic in the x_1x_2 plane. It is assumed that a stress-strain state in which the strain component e_{33} does not depend on all coordinates and the remaining strain components do not depend on the coordinate x_3 is achieved in the medium.

Since the stress-strain state of the FCM layer is doubly periodic, it is sufficient to consider the periodic element of the structure in the form of a parallelogram of periods (primitive cell).

Let ω_1 and ω_2 be the main periods of the structure. Inside the period parallelogram $P_{m,n}$ *(m,n=0,* $\pm 1, \pm 2, \ldots$*)* are k non-intersecting inclusions (fibres) bounded by the contours L_j. The finite simply-connected regions bounded by the contours L_j are denoted by D_j, and the elastic constants of the medium in the regions D_j (fibers) and D (matrix) are denoted by E_j, ν_j and E, ν, respectively (Figure 1).

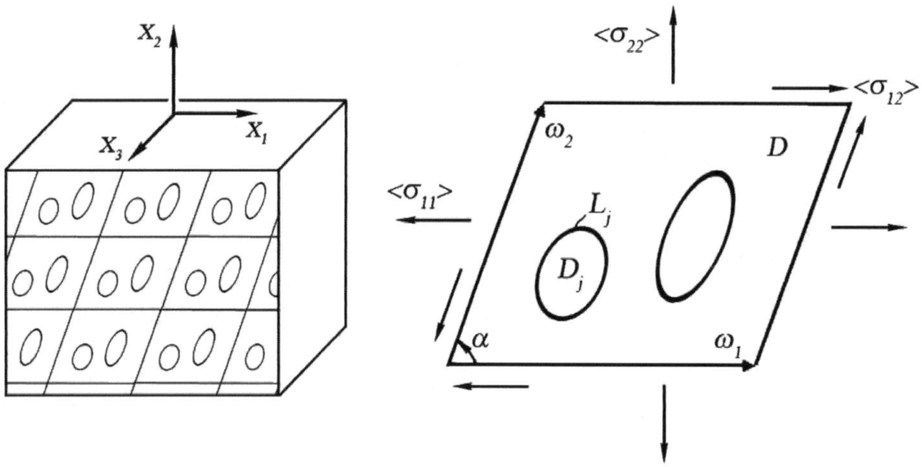

Figure 1. Composite material model

The assumption is made that the fibers are placed in a matrix with some known tension h_j in the x_1x_2 plane. The elastic interaction of the matrix and the fibers is taken to be ideal, which means that the stress and displacement vectors (factoring in the tension) are continuous when passing through L_j.

3. Problem reduction to a system of integral equations

Let Hooke's law be valid in the domain D. Then

$$e_{11} = \frac{1}{E}(\sigma_{11} - \nu\sigma_{22}),\ e_{22} = \frac{1}{E}(\sigma_{22} - \nu\sigma_{11}),\ e_{12} = \frac{1}{\mu}\sigma_{12} \tag{1}$$

The stress equilibrium equations are:

$$\frac{\partial \sigma_{11}}{\partial x_1} + \frac{\partial \sigma_{12}}{\partial x_2} = 0,\quad \frac{\partial \sigma_{12}}{\partial x_1} + \frac{\partial \sigma_{22}}{\partial x_2} = 0 \tag{2}$$

Strain compatibility equations:

$$\frac{\partial^2 e_{11}}{\partial x_2^2} + \frac{\partial^2 e_{22}}{\partial x_1^2} = \frac{\partial^2 e_{12}}{\partial x_1 \partial x_2}. \tag{3}$$

Introducing the stress function (in this case – Airy stress function) with the formulas:

$$\sigma_{11} = \frac{\partial^2 U}{\partial x_2^2}; \sigma_{22} = \frac{\partial^2 U}{\partial x_1^2}; \sigma_{12} = -\frac{\partial^2 U}{\partial x_1 \partial x_2}, \tag{4}$$

to the relations (1) and (3), we obtain the biharmonic equation:

$$\nabla^2 \nabla^2 U(x_1, x_2) = 0, \tag{5}$$

here, the equilibrium equations are satisfied automatically. Thus, the Airy stress function is biharmonic. Introducing the complex variable $z = x_1 + ix_2$ allows expressing any biharmonic function in terms of two arbitrary analytical functions with respect to D (potentials) $\varphi(z)$, $\psi(z)$ according to the Goursat formula [6]. Thus, the stresses and displacements acting in the medium will be written as:

$$\begin{gathered}
\sigma_{11} + \sigma_{22} = 4\mathrm{Re}\Phi(z), \\
\sigma_{11} - \sigma_{22} + 2i\sigma_{12} = 2[\bar{z}\Phi'(z) + \Psi(z)], \\
\sigma_{33} = 2\mu(1+\nu)e_{33} + 4\nu\mathrm{Re}\Phi(z),) \\
2\mu(u_1 + iu_2) = \kappa\varphi(z) - z\overline{\Phi(z)} - \overline{\psi(z)},
\end{gathered} \tag{6}$$

where $\Psi(z) = \psi'(z), \Phi(z) = \varphi'(z)$, $\kappa = 3 - 4\nu$. An overscore above a function means that it is a complex conjugate function.

We will consider stress fields with a symmetry group similar to the region *D*. In this case, the stresses in *D* must have a doubly periodic structure. The composite material plane deformation problem can thus be formulated as follows. Defining the functions $\varphi(z), \psi(z)$ and $\varphi_j(z), \psi_j(z)$, regular, respectively, in the domains *D*, D_j *(j=1,2,..., k)* and satisfying at the boundary surface $L = \bigcup L_j$ the following matrix-fibres coupling conditions:

– stress vector continuity

$$\varphi(t) + t\overline{\Phi}(t) + \overline{\psi}(t) = \varphi_j(t) + t\overline{\Phi}_j(t) + \overline{\psi}_j(t), \tag{7}$$

- displacement vector continuity (factoring in the tension)

$$\frac{1}{\mu}\left[\kappa\varphi(t) - t\overline{\Phi}(t) - \overline{\psi}(t)\right] = \frac{1}{\mu_j}\left[\kappa_j\varphi_j(t) - t\overline{\Phi}_j(t) - \overline{\psi}_j(t)\right] + 2h_j(t). \tag{8}$$

Here,

$$t \in L_j,\ \mu = \frac{E}{2(1+\nu)},\ \mu_j = \frac{E_j}{2(1+\nu_j)},\ \kappa = \frac{3-\nu}{1+\nu},\ \kappa_j = \frac{3-\nu_j}{1+\nu_j}.$$

It is assumed that all the periodicity conditions are satisfied automatically due to a special form of representations of the sought regular functions. As shown in [7], the required functions $\varphi(t)$, $\psi(t)$ can be expressed in terms of two unknown complex functions (densities) *p(t)* and *q(t)*, and such that to determine *p(t)* and *q(t)*, a system of integral equations equivalent to the original boundary value problem is obtained. Substituting the expressions for $\varphi(t)$ and $\psi(t)$ in the coupling conditions (7) (8), we obtain a system of second-order Fredholm integral equations:

$$p(t_0) - M_j(p(t), q(t), t_0) = R_j(t_0),$$
$$q(t_0) - N_j(p(t), q(t), t_0) = Q_j(t_0), \quad (9)$$

Solving the system (9) gives the values of the densities *p(t), q(t)* on the contour *L*. Thereafter, the derivatives of the complex potentials are calculated, followed by the stresses at the component boundaries using the formulas [2]:

$$\sigma_r(t) = \mathrm{Re}[2\Phi(t) - \bar{t}\Phi'(t) - \Psi(t)],$$
$$\sigma_\theta(t) = \mathrm{Re}[2\Phi(t) + \bar{t}\Phi'(t) + \Psi(t)]. \quad (10)$$

4. Influence of residual stresses

The heterogeneous structure of FCMs in power structures is usually formed at elevated temperatures 120...180 ^{0}C with subsequent cooling to ambient room temperature, which ensures the highest mechanical properties. The cooling process is accompanied by thermal deformations of the FCM components. Since the fibres and the matrix have different thermoelastic properties, they undergo constrained deformation proportionally to the difference in their thermal expansion coefficients and the cooling mode temperature range. The separation of the main acting stresses in the cross-section based on polarisation-optical measurement results showed that the plane perpendicular to the direction of fibre-reinforcement experiences radial residual compressive stresses σ_r and tangential tensile stresses σ_θ, which attain their highest values at the boundary surface [8].

For analytical calculation of residual stresses in a uniaxially reinforced composite, an experimentally-derived thermoelastic analogy [8] is used based on which residual stresses are considered thermoelastic, proportionally to the difference in the thermal expansion coefficients of the matrix and the reinforcing fibres and the temperature difference ΔT during composite cooling.

The constrained temperature displacements of the reinforcing fibres and the matrix can then be calculated from the equations:

$$U + \mathrm{i}V = t(\alpha - \alpha_{2j})\Delta T,$$
$$W = t(\alpha - \alpha_{1j})\Delta T, \quad t \in L. \quad (11)$$

Here, α_{1j} and α_{2j} are the thermal expansion coefficients along and across the *j*-th fiber, respectively, and α is the thermal expansion coefficient of the matrix. Axial residual stresses σ_3 develop due to the difference in the Poisson's ratios of the fibres and the matrix.

The residual stresses σ_1 and σ_2 acting in the cross-sections are calculated from the solution of the composite plane deformation problem (9), under the action of the following load:

$$\langle\sigma_{11}\rangle = \langle\sigma_{22}\rangle = \langle\sigma_{12}\rangle = 0, h_j \neq 0. \quad (12)$$

Here, h_j is the lateral fibre tension caused by the thermal deformations of the matrix and the fibre. Its value can be calculated from the equation:

$$h_j(t) = t\Delta T[(\alpha - \alpha_{2j}) + (\alpha - \alpha_{1j})(\nu - \nu_j)]. \quad (13)$$

In contrast to the previously obtained solutions, the transverse deformations of both the matrix and the fibres are taken into account here, i.e. stresses are determined both in the areas occupied by the fibres and in the area occupied by the matrix. In addition, this formulation of the problem allows not only to determine the initial stress-strain state of an FCM corresponding to the loading case (12), but also the effect of residual stresses on the bearing capacity of a composite under an external load.

5. Calculation results

The proposed method was used to carry out calculations of fiberglass having a hexagonal lattice $\omega_1 = 2, \omega_2 = 2e^{i\pi/3}$ and a radius of $R=0.75$ at the inclusion nodes (Figure 2), at a temperature difference $\Delta T = -120^\circ C$.

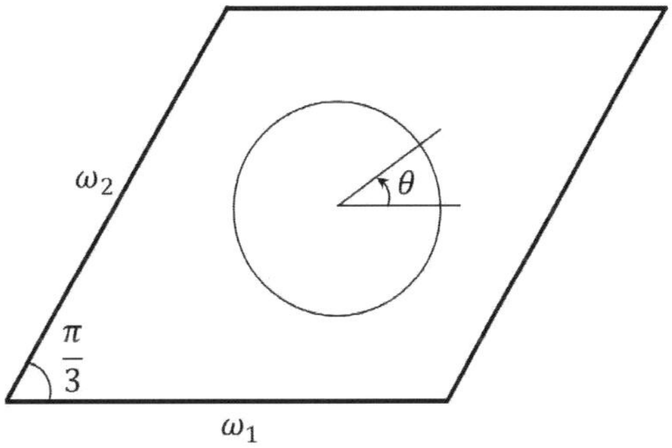

Figure 2. Hexagonal lattice structure

The graphs (Figure 3 and Figure 4) show the distribution of residual radial and tangential stresses for fiberglass along the matrix-fiber interface, where they attain their maximum values.

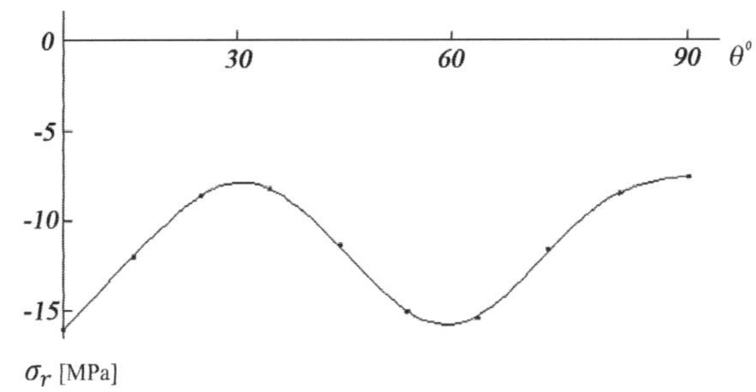

Figure 3. Distribution of radial residual stresses along matrix-fibre interface

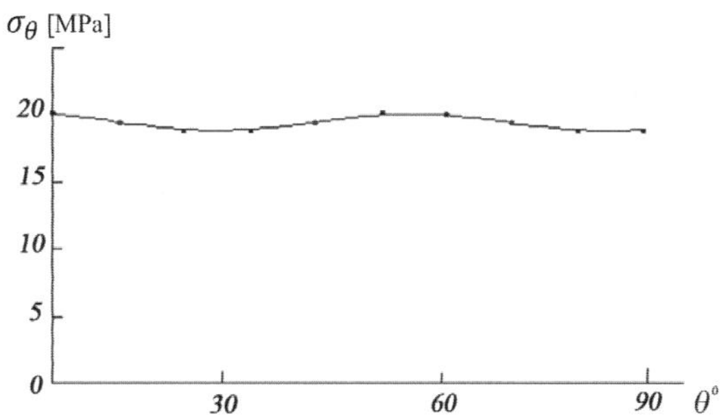

Figure 4. Distribution of tangential residual stresses along matrix-fibre interface

It follows from the calculations that the maximum residual stresses reach 50% of the ultimate strength of the matrix. It is obvious that such a high level of the initial stress state in the matrix will have a significant impact on both the strength of the material and the nature of its destruction under an external load.

To identify the influence of the degree of fibre-reinforcement and the curing temperature on the initial stress state level, graphs of the distribution of relative equivalent residual stresses for fiberglass (Figure 5), calculated using the tensor-invariant Goldenblat-Kopnov strength criterion [10], are drawn.

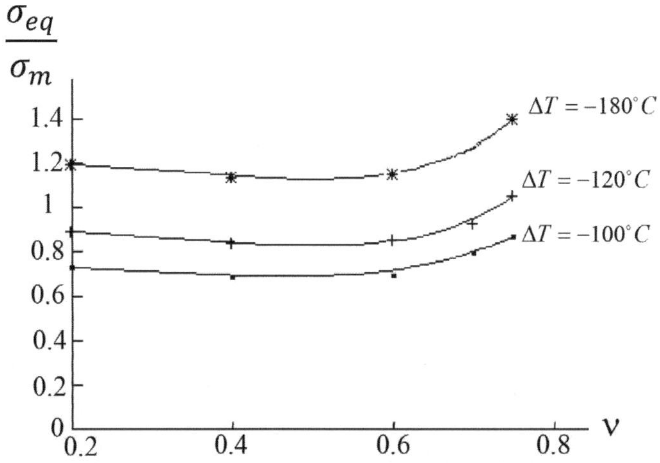

Figure 5. Dependence of relative equivalent residual stresses on the degree of fibre-reinforcement

As can be seen from the graphs, at high degrees of fibre-reinforcement and temperatures, the matrix can break even during fabrication of the material without applying an external load. In addition, the specific optimal degree of fibre-reinforcement at which the residual stresses are minimal for different curing temperatures was determined to be equal to $\nu \approx 0.6$.

6. Conclusion

An analytical method was developed for calculating the initial stress-strain state of a unidirectional fibrous composite material arising from the constrained thermal deformations of the matrix and fibre during its fabrication. Analysis of the initial stress state revealed that at high degrees of reinforcement and curing temperatures of the matrix, the matrix can break even during fabrication of the material without applying an external load. The optimal degree of reinforcement at which residual stresses are minimal for different curing temperatures was determined.

The method proposed can be used to develop an optimal technological process for fabricating a composite and to factor in the influence of the initial stress-strain state of the composite on its load-bearing capacity under an external load.

References

[1] Wildeman V E 2012 Experimental studies of the properties of materials under complex thermomechanical influences. Moscow. Fizmatlit. 204 p. *(in Russian)*

[2] Zarubin V S, Kuvyrkin G N. 2002 Mathematical models of thermomechanics. Moscow. Fizmatlit. 168 p. *(in Russian)*

[3] Zinoviev P A, Kulish G G, Tsvetkov S V 2005 Processes of deformation and fracture of composite materials under high-intensity triaxial loading. Moscow. BMSTU. 94 p. *(in Russian)*

[4] Turusov R A, Memarianfard H 2015 Discrete model in the analysis of residual stresses in unidirectional winding cylinders made of fiber-reinforced plastic during cooling. *Moscow State University of Civil Engineering Bulletin*. **1.** Pp. 27-35.

[5] Memarianfard M., Turusov R A, Memarianfard H 2016 Numerical and experimental studies of the solidity of a thick-walled anisotropic shell. *Moscow State University of Civil Engineering Bulletin*. **7.** Pp. 36-45

[6] Muskhelishvili N I 1966 Some basic problems of the mathematical theory of elasticity. Moscow. Nauka. 708 p. *(in Russian)*

[7] Grigolyuk E I, Filshtinsky L A 1992 Periodic piecewise homogeneous elastic structures. Moscow. Nauka. 287p. *(in Russian)*
[8] Molodtsov G A 1993 Stressed elements of aircraft structures made of composite materials. Moscow. Mashinostoenie. 224 p. *(in Russian)*
[9] Karpov V V, Semenov A A 2015 Strength criteria for thin-walled orthotropic shells. Part 2: Calculations and analysis. *Civil Engineers Bulletin.* 1 **(48).** Pp. 60-70.

Flexural vibrations of the console with frictio

© A.A. Pozhalostin[1], A.V. Panshina[1]

[1]Bauman Moscow State Technical University, Moscow, 105005, Russia

E-mail: a.pozhalostin@mail.ru panalv@mail.ru

Abstract. An approximate approach is considered for determining the logarithmic decrement of Flexural vibrations of an elastic console with a load at the end. The method proposed in Postnikov V. S.'s monograph on internal friction in metal is used. In this problem statement, the friction is assumed to be small. Therefore, the forms of natural vibrations of the console with friction are considered equal to the forms of natural vibrations of the console without friction. An approximate solution of this problem is obtained using a frequency experiment.

1. Introduction.

The paper presents an approach based on the idea expressed by Coulomb at one time. A method for determining the logarithmic decrement of Flexural vibrations of an elastic console with friction, proposed in V. S. Postnikov's monograph on internal friction in metals [1], is used. Bending vibrations of an elastic homogeneous beam with a load at the end are considered. An approximate solution of this problem is obtained using a frequency experiment.

2. Setting and solving the problem.

The oscillating system (Fig. 1) consists of a pinched homogeneous beam of linear density μ_0 [2,3]. The length of the beam is $OA = l$. At the free end of the beam is a point weight of mass m.

We assume the following assumptions: the beam is perfectly elastic, the vibrations are small, the bend is straight, and the internal friction forces are proportional to the strain rate in the first degree [1,4].

Let $y(x,t)$ be the bending of a beam (figure 1). The differential equation of small bending vibrations of an ideally elastic homogeneous beam has the form [1,3]:

$$EJ_0 \frac{\partial^4 y}{\partial x^4} + \mu_0 \frac{\partial^2 y}{\partial t^2} = 0, \qquad (1)$$

where EJ_0 is the flexural stiffness [2], E - Jung 's modulus of the first kind.

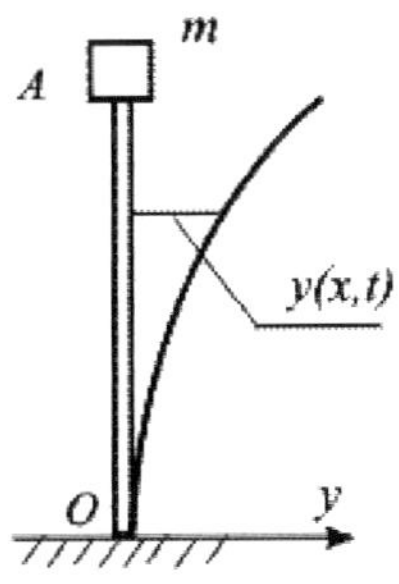

Figure 1. Diagram of a pinched homogeneous beam with a point load at the free end.

Note that $\frac{\partial^2 y}{\partial t^2} = -\omega^2 y$, where ω is the frequency of free vibrations of the system.

Let's introduce the designation $\frac{\mu_0 \omega^2}{EJ_0} = \lambda^4$. Here λ is the eigenvalue of our boundary value problem.

Then equation (1) will take the form:

$$\frac{\partial^4 y}{\partial x^4} - \lambda^4 y = 0 \qquad (2)$$

Looking for the deflection of the beam in the following form [3-5]:

$$y = f(x) \cdot s(t). \qquad (3)$$

Since the function $s(t) \neq 0$, we can write the differential equation for the oscillation form $f(x)$:

$$f^{IV} - \lambda^4 f = 0, \qquad (4)$$

where $f^{IV} = \frac{d^4 f}{dx^4}$.

The boundary conditions of the problem for $x = 0$ have the form [5]:

$$y(0,t) = 0; \ \frac{\partial y}{\partial x}(0,t) = 0. \qquad (5)$$

We assume the boundary conditions for $x = l$ as follows:

$$M_{flex}(l,t) = 0, \quad m\frac{d^2 y}{dt^2} + Q(l,t) = 0. \qquad (6)$$

The bending moment and cutting force (figure 2) have the form:

$$M_{flex}(x,t) = -EJ_0 \frac{\partial^2 y}{\partial x^2}, \quad Q(x,t) = -EJ_0 \frac{\partial^3 y}{\partial x^3}. \qquad (7)$$

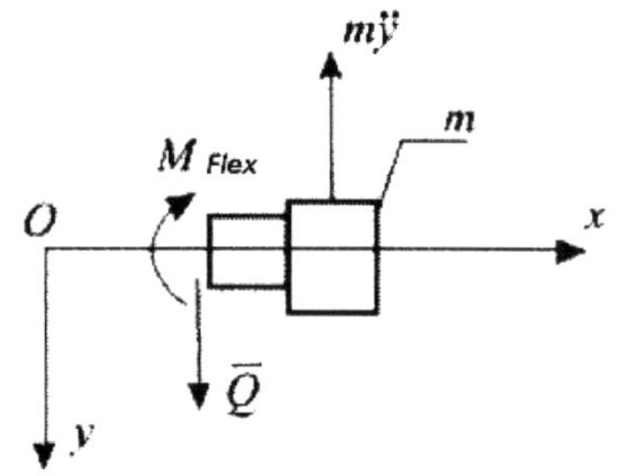

Figure 2. Diagram of a beam element with a point load with forces and moments acting on it in accordance with the D'Alembert principle.

Using the D'Alembert principle, we obtain the equation:

$$-m\ddot{y}+Q(l,t)=0\,. \qquad (8)$$

Since $\ddot{y}=-\omega^2 y$ [6], equation (8) takes the form:

$$m\omega^2 y(l,t)-EJ_0\frac{\partial^3 y(l,t)}{\partial x^3}=0\,. \qquad (9)$$

Let's write out the solution of the differential equation (4) for the oscillation form $f(x)$ [4,5]:

$$f(x)=C_1e^{\lambda x}+C_2e^{-\lambda x}+C_3\cos\lambda x+C_4\sin\lambda x\,. \qquad (10)$$

The integration constants C_i $(i=1,2,3,4)$ are determined from the boundary conditions (5) and (6).

It is known [6] that the solution of differential equations for a linear vibrational system with two degrees of freedom is reduced to the solution of a homogeneous system of two algebraic equations:

$$\begin{cases} C_1a_{11}+C_2a_{12}=0 \\ C_1a_{21}+C_2a_{22}=0 \end{cases}$$

A system has a non-zero solution if its determinant is zero:

$$\Delta(\lambda)=a_{11}a_{22}-a_{12}a_{21}=0\,. \qquad (11)$$

Equation (11) is the frequency transcendental equation of the problem.

Substitute the desired solution (10) instead of $y(x,t)$ in the boundary conditions (5) and (6), taking into account expressions (7) for $M_{flex}(l,t)$ and $Q(l,t)$. Let's write out the determinant of the obtained algebraic equations and equate it to zero. After simple transformations, we obtain the frequency equation:

$$\begin{gathered}(ch\lambda l-\cos\lambda l)\cdot[m\omega^2(sh\lambda l-\sin\lambda l)-EJ_0\lambda(ch\lambda l-\cos\lambda l)]- \\ -(sh\lambda l+\sin\lambda l)\cdot[m\omega^2(ch\lambda l-\cos\lambda l)-EJ_0\lambda(sh\lambda l+\sin\lambda l)]=0\end{gathered}\,. \qquad (12)$$

Solving the transcendental equation (12), we can calculate the values of λ_i and the natural oscillation frequency ω_i.

Based on [1,4], we believe that the internal friction in the rod material and in the atmosphere surrounding the installation is approximately directly proportional to the rate of change in the deformation in the beam and the speed of movement of the material elements of the surrounding atmosphere. Therefore, when calculating the viscous friction forces, we use the Rayleigh formalism [2]: $\Phi=\dfrac{\mu v^2}{2}$, where $v=\dfrac{\partial y(x,t)}{\partial t}$ is the velocity of the material point of the system.

In the future, we will only take into account the first tone of the system's vibrations. Therefore [4,6], we have a differential equation for $s(t)$:

$$\ddot{s} + 2n\dot{s} + \omega^2 s = 0 .$$

Here the natural frequency ω of the first tone of the transverse vibrations of the system without taking into account the forces of viscous friction is the solution of equation (12).

We denote the experimental values of the frequency of the first tone (taking into account viscous friction) by $\omega_{1\exp}$.

Since $\omega_{1\exp}{}^2 = \omega^2 - n^2$ [3-5] and $\omega_{1\exp} = \frac{2\pi}{T_1}$, $\omega = \frac{2\pi}{T}$, we have:

$$n^2 = \left(\frac{2\pi}{T}\right)^2 - \left(\frac{2\pi}{T_1}\right)^2 .$$

Let us now write out the logarithmic decrement of the system's oscillations [7]:

$$\eta = nT_1 = T_1\sqrt{\left(\frac{2\pi}{T}\right)^2 - \left(\frac{2\pi}{T_1}\right)^2} .$$

Here T_1, $\omega_{1\exp}$ are the period and frequency of the first tone of the system's vibrations, determined experimentally.

3. Conclusion.

An approximate solution to the problem of determining the logarithmic decrement of Flexural vibrations of an elastic console with friction using a frequency experiment is obtained. The validity of this approach can be tested experimentally using a simple installation with a pinched beam.

Literature.

[1] Postnikov V S 1974 *Internal friction in metals* (Moscow: Metallurgy) 352 p
[2] Feodosyev V I 2016 *Material Resistance* (Moscow: Bauman Moscow state technical University) 543 p
[3] Strelkov S P 2019 *Introduction to the Theory of Vibrations* (St. Petersburg: LAN) 560 p
[4] Kolesnikov K S 2003 *Dynamics of rockets* (Moscow: Mashinostroenie) 519 p
[5] Ilyin M M and Kolesnikov K S and Saratov Y S 2003 *Theory of Vibrations* (Moscow: Publ. House of the Bauman Moscow State Techn. University) 272 p
[6] Kolesnikov K S and Dubinin V V and others 2017 *Course of theoretical mechanics* ((Moscow: Publ. House of the Bauman Moscow State Techn. University) 584 p
[7] Coulomb C A 1821 *Theory of simple machines* (Paris) 388

IOP Conf. Series: Materials Science and Engineering 1191 (2021) 012011 doi:10.1088/1757-899X/1191/1/012011

The equations of a solid body motion

V V Lapshin

Bauman Moscow State Technical University, 2-ya Baumanskaya st.,5, Moscow, Russia

E-mail: lapshin032@bmstu.ru

Abstract. Traditionally, the equations of motion of a solid body uses the theorem of the center of mass motion and the theorem of change of angular momentum with respect to the center of mass (Koenig's axes). However, in the investigation of a solid body motion in a number of cases, the generalized coordinates are advisable to choose coordinates of another point of a body (pole) does not match the center of mass of the body, and the angles of rotation around this point. We obtain the equations of a solid body motion for these generalized coordinates. This is important for mathematical modeling of the motion of various objects, including vehicles. During the motion of a vehicle, its mass-inertial characteristics may change, including the position of the center of mass and the direction of the main axes of inertia, body mass and moments of inertia.

1. Introduction

Usually, to describe a solid body motion, the theorem of the center of mass motion and the theorem on the change in the angular momentum relative to the center of mass (Koenig's axes) are used [1-7]. In this case, the coordinates of the center of mass and three angles of rotation of the body relative to the Koenig axes (Euler angles, Krylov angles, aircraft angles or rocket angles) are used as generalized coordinates that determine the position of a rigid body.

However, in the investigation of motion of various moving objects (aircraft, ship, walking apparatus [8-9], etc.), it is advisable to choose coordinates of another point of a body (pole) does not match the center of mass of the body, and the angles of rotation around this point.

This can be due to various reasons. May be the movement of this very point is of particular interest in the investigation of the dynamics of a solid body motion. A solid body is often viewed as a vehicle motion model. In this case, in the process of motion, the mass-inertial characteristics of the body motion can changes including the center of mass position and the direction of the main axes of inertia, mass and moments of inertia. They can be changes either at the same time during loading (unloading) of the vehicle, or during movement as the result of fuel combustion, icing, and dirt adhesion. In the latter cases, we can assume that the mass-inertial characteristics change slowly, and their derivatives can be considered equal to zero. When simulating the dynamics of a solid body, the origin of the moving coordinate system (pole) associated with the solid body and the direction of the axes remain unchanged. These are the pole and axles used in the vehicle (solid body) motion control system.

In this work, the equations of a solid body motion in these generalized coordinates are obtained in the projection on the axis of a moving coordinate system associated with the body.

IOP Conf. Series: Materials Science and Engineering 1191 (2021) 012011 doi:10.1088/1757-899X/1191/1/012011

2. Motion equations of a free solid body

Consider the motion of a solid body (Figure 1) relative to an inertial (stationary) coordinate system $O\xi\eta\zeta$. A movable coordinate system $Axyz$ is rigidly connected to a solid body. Body mass center point C. Let us denote by m the mass of the body, m_k – the mass of the k-th point of the body, $\mathbf{J}$ – the tensor of inertia in a moving coordinate system, $\overline{v}_A$ – the velocity of the pole A, $\overline{v}_C$ – the velocity of the center of mass C, $\overline{\omega}$ – the angular velocity of the rigid body, $\overline{F}$ and $\overline{M} = \overline{M}_A$ is the principal vector and the principal moment relative to the pole A of the forces applied to the solid body. The position of each point of the solid body relative to the fixed coordinate system $O\xi\eta\zeta$ is determined by the radius vector $\overline{r}$, and relative to the moving coordinate system $Axyz$ by the radius vector $\overline{\rho}$. Including for the k-th point of the rigid body we have $\overline{r}_k$ and $\overline{\rho}_k$, for the center of mass C – $\overline{r}_C$ and $\overline{\rho}_C$, for the pole A – $\overline{r}_A$. Moreover, the vectors $\overline{\rho}_C$ and $\overline{\rho}_k$ are constants.

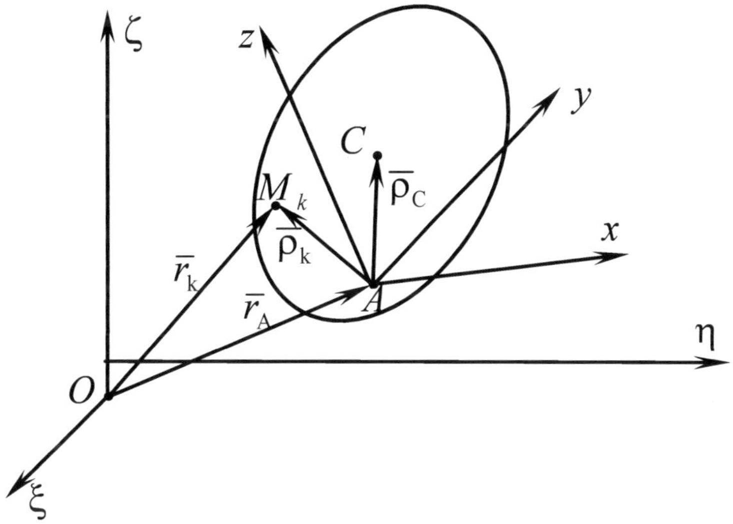

Figure 1.

$$\mathbf{J} = \begin{pmatrix} J_x & -J_{xy} & -J_{xz} \\ -J_{xy} & J_y & -J_{yz} \\ -J_{xz} & -J_{yz} & J_z \end{pmatrix} .$$

For any vector $\overline{b}$, we call the total derivative

$$\frac{d\overline{b}}{dt} = \dot{\overline{b}} \quad ,$$

the derivative calculated in a fixed coordinate system $O\xi\eta\zeta$.

And the local derivative is called

$$\frac{\widetilde{d\overline{b}}}{dt} = \overset{\circ}{\overline{b}} \quad ,$$

derivative calculated in a moving coordinate system $Axyz$.
The total and local derivatives are related by the Boer formula

$$\frac{d\overline{b}}{dt} = \frac{\widetilde{d\overline{b}}}{dt} + \overline{\omega} \times \overline{b} \quad \text{, or} \quad \dot{\overline{b}} = \overset{\circ}{\overline{b}} + \overline{\omega} \times \overline{b} \quad . \tag{1}$$

Note that the total and local derivatives of the angular velocity vector coincide

$$\dot{\overline{\omega}} = \overset{\circ}{\overline{\omega}} \quad . \tag{2}$$

According to Euler's formula

$$\overline{v}_C = \overline{v}_A + \overline{\omega} \times \overline{\rho}_C \quad . \tag{3}$$

The theorem on the motion of the center of mass $m\dot{\overline{v}}_C = \overline{F}$ taking into account (1)–(3) has the form

$$m\left(\overset{\circ}{\overline{v}}_A + \overline{\omega} \times \overline{v}_A + \overset{\circ}{\overline{\omega}} \times \overline{\rho}_C + \overline{\omega} \times \left[\overline{\omega} \times \overline{\rho}_C \right] \right) = \overline{F} \quad . \tag{4}$$

The angular momentum vector relative to a fixed point O is

$$\overline{K}_O = \sum_k \overline{r}_k \times m_k \overline{v}_k = \sum_k \left(\overline{r}_A + \overline{\rho}_k \right) \times m_k \left(\overline{v}_A + \overline{\omega} \times \overline{\rho}_k \right) \quad .$$

Then by virtue of relation (3), and

$$\sum_k m_k = m \quad , \quad \sum_k m_k \overline{\rho}_k = m \overline{\rho}_C \quad , \quad \sum_k m_k \overline{\rho}_k \times \left[\overline{\omega} \times \overline{\rho}_k \right] = \mathbf{J} \overline{\omega} \quad ,$$

we obtain

$$\overline{K}_O = \overline{r}_A \times m \overline{v}_C + m \overline{\rho}_C \times \overline{v}_A + \mathbf{J} \overline{\omega} \quad . \tag{5}$$

Theorem on the change in angular momentum relative to a fixed point O

$$\dot{\overline{K}}_O = \overline{M}_O = \sum_i \overline{r}_i \times \overline{F}_i = \sum_i \left(\overline{r}_A + \overline{\rho}_i \right) \times \overline{F}_i = \overline{r}_A \times \overline{F} + \overline{M}_A = \overline{r}_A \times m \dot{\overline{v}}_C + \overline{M} \quad .$$

Then by virtue of relation (1), (5) and the fact that the tensor of inertia in the moving coordinate system $Axyz$ is constant, has the form

$$\overline{r}_A \times m \dot{\overline{v}}_C + \overline{v}_A \times m \overline{v}_C + m \left[\overline{\omega} \times \overline{\rho}_C \right] \times \overline{v}_A + m \overline{\rho}_C \times \left(\overset{\circ}{\overline{v}}_A + \overline{\omega} \times \overline{v}_A \right) +$$
$$+ \mathbf{J} \dot{\overline{\omega}} + \overline{\omega} \times \mathbf{J} \overline{\omega} = \overline{r}_A \times m \dot{\overline{v}}_C + \overline{M} \quad .$$

Canceling identical terms on the left and right sides and then by virtue of virtue of (3)

$$\overline{v}_A \times m \overline{v}_C + m \left[\overline{\omega} \times \overline{\rho}_C \right] \times \overline{v}_A = \overline{v}_A \times m \left(\overline{v}_A + \overline{\omega} \times \overline{\rho}_C \right) + m \left[\overline{\omega} \times \overline{\rho}_C \right] \times \overline{v}_A =$$
$$= m \overline{v}_A \times \left[\overline{\omega} \times \overline{\rho}_C \right] + m \left[\overline{\omega} \times \overline{\rho}_C \right] \times \overline{v}_A = 0 \quad .$$

IOP Conf. Series: Materials Science and Engineering 1191 (2021) 012011 doi:10.1088/1757-899X/1191/1/012011

Then by virtue of (2), we obtain

$$\mathbf{J}\overset{\circ}{\overline{\omega}}+\overline{\omega}\times\mathbf{J}\overline{\omega}+m\overline{\rho}_C\times(\overset{\circ}{\overline{v}}_A+\overline{\omega}\times\overline{v}_A)=\overline{M} \quad . \tag{6}$$

Equations (4), (6) must be supplemented with Euler's kinematic equations and equations $\dot{\overline{r}}_A=\overline{v}_A$. Note that in these equations $\overline{F}$ and $\overline{M}$ are functions of time, coordinates, and velocities of the pole A, angular coordinates and angular velocity of a solid body.

In the case when the pole A coincides with the center of mass C, the vector $\overline{\rho}_C=0$ and equations (4), (6) have the traditional form of the theorem of the center of mass motion and the theorem on the change in angular momentum with respect to the Koenig axes. In this case, it is advisable to write equations (4) in projection on the axis of a fixed coordinate system. As a result, we have

$$m\dot{\overline{v}}_C=\overline{F} \quad , \quad \mathbf{J}\overset{\circ}{\overline{\omega}}+\overline{\omega}\times\mathbf{J}\overline{\omega}=\overline{M} \quad . \tag{7}$$

Equations (7) are much simpler than equations (4), (6). Nevertheless, when modeling the dynamics of movement and constructing algorithms for controlling the movement of various vehicles, it is advisable to use equations (4), (6) because they allow taking into account changes in the mass-inertial characteristics of an object during its movement.

Equations (4), (6), written in the projection on the axis of the moving coordinate system $Axyz$, have the form

$$\begin{aligned}
&m[\overset{\circ}{v}_x+\overset{\circ}{\omega}_y z_C-\overset{\circ}{\omega}_z y_C+\omega_y v_z-\omega_z v_y+\omega_y(\omega_x y_C-\omega_y x_C)-\omega_z(\omega_z x_C-\omega_x z_C)]=F_x \quad ,\\
&m[\overset{\circ}{v}_y+\overset{\circ}{\omega}_z x_C-\overset{\circ}{\omega}_x z_C+\omega_z v_x-\omega_x v_z-\omega_x(\omega_x y_C-\omega_y x_C)+\omega_z(\omega_y z_C-\omega_z y_C)]=F_y \quad ,\\
&m[\overset{\circ}{v}_z+\overset{\circ}{\omega}_x y_C-\overset{\circ}{\omega}_y x_C+\omega_x v_y-\omega_y v_x+\omega_x(\omega_z x_C-\omega_x z_C)-\omega_y(\omega_y z_C-\omega_z y_C)]=F_z \quad ,\\
&J_x\overset{\circ}{\omega}_x-J_{xy}\overset{\circ}{\omega}_y-J_{xz}\overset{\circ}{\omega}_z+(J_z-J_y)\omega_y\omega_z+J_{yz}(\omega_z^2-\omega_y^2)+\omega_x(J_{xy}\omega_z-J_{xz}\omega_y)+\\
&+m[\overset{\circ}{v}_z y_C-\overset{\circ}{v}_y z_C+y_C(\omega_x v_y-\omega_y v_x)-z_C(\omega_z v_x-\omega_x v_z)]=M_x \quad ,\\
&J_y\overset{\circ}{\omega}_y-J_{xy}\overset{\circ}{\omega}_x-J_{yz}\overset{\circ}{\omega}_z+(J_x-J_z)\omega_x\omega_z+J_{xz}(\omega_x^2-\omega_z^2)+\omega_y(J_{yz}\omega_x-J_{xy}\omega_z)+\\
&+m[\overset{\circ}{v}_x z_C-\overset{\circ}{v}_z x_C-x_C(\omega_x v_y-\omega_y v_x)+z_C(\omega_y v_z-\omega_z v_y)]=M_y \quad ,\\
&J_z\overset{\circ}{\omega}_z-J_{xz}\overset{\circ}{\omega}_x-J_{yz}\overset{\circ}{\omega}_y+(J_y-J_x)\omega_y\omega_z+J_{xy}(\omega_y^2-\omega_x^2)+\omega_z(J_{xz}\omega_y-J_{yz}\omega_x)+\\
&+m[\overset{\circ}{v}_y x_C-\overset{\circ}{v}_x y_C+x_C(\omega_z v_x-\omega_x v_z)-y_C(\omega_z v_y-\omega_y v_z)]=M_z \quad .
\end{aligned} \tag{8}$$

3. Plane motion of a solid body

In the plane motion of a solid body, when all points of the body move parallel to the plane $O\xi\eta$ of the fixed coordinate system, which coincides with the plane Axy of the moving coordinate system. Without loss of generality, we can assume, that the center of mass located in the same plane. Then, $z_C=0,\ \omega_x=\omega_y=0,\ \omega_z=\omega,\ v_z=0$ and equations of motion (8) have the form

$$m(\mathring{v}_x - \mathring{\omega} y_C - \omega v_y - \omega^2 x_C) = F_x \quad ,$$

$$m(\mathring{v}_y + \mathring{\omega} x_C + \omega v_x - \omega^2 y_C) = F_y \quad ,$$

$$J_z \mathring{\omega} + m(\mathring{v}_y x_C - \mathring{v}_x y_C + x_C \omega v_x - y_C \omega v_y) = M_z \quad .$$

4. Conclusion

When simulating the dynamics of a solid body motion in several cases (in particular, when simulating the motion of bodies, whose mass-inertial characteristics can change during the motion), it is advisable to use the coordinate of a certain characteristic point of the body (pole) and the angles of rotation of the body around this point as generalized coordinates. In this work, the equations of motion of the body in these generalized coordinates are obtained.

References

[1] Appell P 1953 *Traité de Mécanique Rationnelle, Tome 2, Dynamique des Systèmes, Mécanique Analitique* (Paris: Gauther Villars)

[2] Bolotin S V, Karapetian A V, Kugushev E I, Treschev D V 2010 *Theoretical Mechanics* (Moscow: Moscow State University Press) *[in Russian]*

[3] Butenin N V, Lunts Ia L, Merkin D R 2009 *Course of Theoretical Mechanics* (St. Petersburg: Izdatel'stvo Lan') *[in Russian]*

[4] Golubev Yu Ph 2019 *Foundations of Theoretical Mechanics* foundations (Moscow: Moscow State University Press) *[in Russian]*

[5] Kolesnikov K S and etl. 2017 *Course of Theoretical Mechanics* (Moscow: Bauman Moscow State Technical University Press) *[in Russian]*

[7] Markeev A P 2001 *Theoretical Mechanics* (Moscow, Igevsk: Regular and Chaotic Dynamics)

[8] Okhotsimsky D E and Golubev Yu Ph 1980 *Mechanics and Motion Control of Automatic Walking Vehicle* (Moscow: Science) *[in Russian]*

[9] Lapshin V V 2012 *Mecanics and Motion Control of Walking Machines* (Moscow: Bauman Moscow State Technical University Press) *[in Russian]*

Models of visual information processing systems

Shkapov P.M.[1], **Blagoveshchensky I.G.**[1]
Blagoveshchensky I.G.[2],

[1]Bauman Moscow State Technical University, Moscow, 105005 Russia
[2]Moscow State University of Food Production, 11 Volokolamskoe highway, Moscow, 125080, Russia

E-mail: spm@bmstu.ru; igblagov@mgupp.ru

Abstract. The article shows the importance of solving various production problems using processing, analysis and display of visual information. The existing systems for processing visual information are considered and the insufficiently developed use of automated image processing systems in production systems that solve rather complex and "intelligent" problems of image analysis is noted. The prospects of using digital modeling of object images are shown. The article provides a literary review of works on the development of methods and models for effective processing of visual control information, which showed that there are many unresolved issues. This work is aimed at solving some of them. The relevance of the topic and the objectives of the research are determined. Based on the results of the studies performed, a model of the visual information processing system was developed. From the object under study, a video stream arrives at the input of the model of the visual information processing system, which is first selected as a sequence of frames (frames) of a fixed dimension. Then there is a preliminary processing of the information received from the object, its segmentation, extraction of features and decision making. Here, the traditional number of homogeneous segments can be applied, into which the object is divided using the principle of maximum posterior probability. The result obtained is then used in the adaptation block. The proposed model of the visual information processing system was implemented in the Qt Creator environment in C ++.

1. Introduction

The constant growth of requirements for the quality of manufactured products (detection of defects, analysis and prediction of the properties of finished products, structural analysis of the materials obtained) increases the labor intensity of its control, as a result of which there is a need to automate this process using computer vision systems [1].

Also, in various fields of science and technology, there is an increase in the need for the analysis and processing of visual information, for example, in vision systems for robots and flexible automated production, during product orientation, real-time analysis of quality indicators of various types of products (color, shape, texture, sizes, etc.) [2, 3]. An important class of this line of tasks is visual information processing systems. Among them, the most widely used are automated systems for processing visual information [4]. Problem-oriented applied automated systems for processing visual information use, as a rule, a fairly rich arsenal of various methods and data processing tools, analyze various input information, and actively use human capabilities.

Therefore, such systems contain various functional blocks and subsystems that provide not only the solution of the actual problems of image processing, but also the input and processing of other types of data and knowledge, the accumulation, storage and display of results, the organization of interaction with the user and other systems. The presence of heterogeneous components working together to achieve a common goal is a sign of the transformation of modern automated systems for processing visual information into integrated systems [5]. Image analysis work becomes a kind of repeated experimentation with the original data. A characteristic feature of such an experiment is the combined use of image processing methods, mathematical statistics, artificial intelligence, data analysis, and other methods [6, 7]. Moreover, as far as possible, all available data are entered into processing, including intermediate results. There is a need to organize effective data exchange in the system, create advanced storage facilities and provide access to data (including video databases) [8, 9]. At present, this is already typical for many automated visual information processing systems (ASOVI) used in remote sensing [10]. The key element of such systems is usually considered to be a processor that processes visual information. Currently, workstations for image processing and systems based on personal computers are widely used [11, 12].

Analysis of existing methods of processing visual information revealed the following disadvantages:

1. Lack of clarity of parameters, regardless of the used method of processing visual information. One of the ways to solve this problem is provided by the Feature Analyst system [13], which implements programs such as ArcGIS, IMAGINE, GeoMedia, etc. as a plug-in into the data processing system.

A model linking low-level features with the characteristics of significant objects is automatically created and iteratively refined based on user feedback. A set of machine learning algorithms implemented in the system (artificial neural networks, decision trees, Bayesian networks, k-nearest neighbor algorithm) are adjusted to the selection of significant objects. This allows to significantly simplify the user interface, but at the same time partially hides the image analysis mechanism from it, which does not allow to correctly interpret the results and adjust the parameters of their obtaining.

2. When processing visual information, not only spectral and textural characteristics should be taken into account, but also the relative position of objects in the image, which plays an important role in making final decisions by the operator.

3. The processing of high-resolution visual information requires a classification at a higher level compared to the level of the original image. In high-resolution images, there may be significant variations in the characteristics of individual pixels in the composition of the same object, which can lead to incorrect classification. Aggregated processing of image elements assumes image segmentation according to spectral and texture characteristics. Further, the classification is performed on the obtained segments. The classification takes into account the hierarchy of segments, their spectral characteristics, shape characteristics, spatial relationships between them.

In the early 2000s, object-oriented image analysis (the OBIA Object-Based Image Analysis approach) was intensively developed [14, 15]. By 2014, the disadvantages of this approach appeared due to the hiding of the mechanisms for selecting objects, which does not contribute to the improvement of the operator's qualifications and limits him in this endeavor. The operator forms a rather critical attitude towards automation tools [16].

In [17, 18], the analysis of the structure and phase composition is carried out by processing images of microstructures using the basic principles of stereometry and the theory of pattern recognition. The currently existing methods of automated analysis of the structure are not complex, since they are intended only for the structural analysis of the objects under study and do not consider the relationship between its composition and structure. The problem of automatic identification and reconstruction of the shapes of various structural elements based on the system analysis and classification of these elements has been little studied [19, 20]. The issues of image processing and recognition are given attention in the works of M.M. Blagoveshchenskaya, Yu. I. Zhuravlev, I.B. Gurevich, R. Gonzalez, R. Woods, A.N. Petryakov, K. Fu and others [21 - 24]. A significant contribution to the problem of processing visual information was made by S. M. Davis, D. A. Landgrebe, T. L. Phillips, F. H. Svein, R. M. Hoffer, Yu. R. A. Shovengerdt and others [25-27].

However, despite the large number of published works, there are still unresolved problems in this area, primarily those related to the development of models of visual information processing systems. Therefore, the subject of this study is methods and algorithms for constructing image models that reflect their content. The object of research is the images of two-dimensional and three-dimensional objects. Purpose of the work: increasing the reliability of automatic processing of visual information of direct and indirect signs of the objects under study through the development of algorithms for machine construction of a formalized description of images.

To achieve this goal, this study sets the following tasks:

1. Research and analysis of existing methods and algorithms for describing images and methods of processing visual information.

2. Development of a model of a visual information processing system.

3. Development of an algorithm for forming the border of a textured object of visual information processing, which differs in the method of detailing a convex polygon.

2. Research

As noted in the introduction, one of the means of increasing the efficiency of the functioning of technical systems using visual data is the use of an automatic system for processing visual information [28]. A person is able to operate with both low-level characteristics of an image (color, texture, shape) and high-level concepts (object, feature, action), while only low-level characteristics are available to a computer. This contradiction leads to the need to develop methods for converting information about a set of raster points into a formalized description of the image, presented in terms of logic and natural language. In view of the need to take into account many interrelated factors, full automation of the visual information processing process seems difficult and requires the development of a model of a visual information processing system.

To obtain such a model, a comprehensive analysis of a set of direct and indirect signs of processing visual information and spatial relationships between the objects under study was carried out with subsequent logical conclusions, which allows for a better interpretation of the data obtained. The main stages of processing visual information include the selection and identification of research objects, carried out by identifying their characteristic combinations of image features - direct and indirect features. Direct signs include the tone and color of the image, contrast, shape and size (area) of the object, geometric parameters of the object's shadow, texture.

The uniformity of direct signs within the study area is a sign of its belonging to the same type. The manifestation of direct signs in the images is due to the physical properties of the depicted objects, as well as technical factors: the scale and quality of shooting, the resolution of the image, etc.

Direct signs allow processing only a part of the specified properties of the objects under study [29]. Difficulties in the processing of visual information by direct signs are associated with their variability depending on the shooting conditions and state of objects and with the similarity of the image in the images of different types of objects [30]. Identical combinations of direct features can correspond to different objects, and objects of the same class in different conditions can be characterized by different sets of features. In the absence of an unambiguous solution, as well as when processing visual information from objects hidden by shadows, indirect signs of analysis and processing are taken into account: relative location, interdependence of objects.

The following generalized stages of the visual information processing process were identified [31]:

1. Classification of objects - assignment of the spatial objects selected in the image to certain classes.

2. Counting the number of objects of each class in the picture.

3. Determination of the geometric characteristics of objects: length, area, height, volume.

4. Determination of the contours of objects and graphic design of the results of processing visual information.

The results of processing visual information are largely determined by the subjective decisions of the operator and his qualifications. Subjectivity was originally incorporated into this process, due to the individual characteristics of the perception of color and tone and the interpretation of the received visual information by each person. Automation makes it possible to provide equal criteria for identifying objects and assessing their characteristics on images obtained at different times.

The first stage is the most difficult for automation due to the need to take into account many interrelated factors. Therefore, the creation of such automated control systems requires the development of methods for their identification, modeling of processes occurring during the processing of visual information, the creation of operational control algorithms implemented by software and hardware systems.

In this study, a model of a visual information processing system has been developed and investigated. It is based on the LCQ color system containing components. L - brightness, C - saturation, Q - hue. The transition from RGB to LCQ color system is done through XYZ color system

$$L = 295{,}8\,/(Y/YD) - 40{,}8,$$

$$C = 380{,}6\sqrt{6{,}25\left[f\left(\frac{X}{X_D}\right) - f\left(\frac{1}{Y_D}\right)\right]^2 + \left[f\left(\frac{Y}{Y_D}\right) - f\left(\frac{Z}{Z_D}\right)\right]^2},$$

$$Q = \frac{128 \cdot \mathrm{arctg}\left(\frac{2 \cdot [f(Y/Y_D) - f(Z/Z_D)]}{5 \cdot [f(X/X_D) - f(Y/Y_D)]}\right)}{\pi};$$

$$f(s) = \begin{cases} \sqrt[3]{s}, & s > 0{,}008856, \\ 7.787s + \frac{16}{116}, & s \le 0{,}008856; \end{cases} \quad \begin{bmatrix} X \\ Y \\ Z \end{bmatrix} = \mathbf{M} \cdot \begin{bmatrix} \Gamma(R) \\ \Gamma(G) \\ \Gamma(B) \end{bmatrix},$$

where M - is a 3x3 transformation matrix from RGB and XYZ systems;
Г (t) - gamma correction function;
XD, YD, ZD - coordinates of the reference white color.

The L component determines the pixel brightness and does not affect the color saturation, in contrast to the weighted sum of the R, G, and B components. Based on it, a histogram of brightness was built, brightness segmentation of visual information was performed, brightness features were calculated, etc.

The Q component tone palette includes primary colors from red to purple. Based on it, a histogram of tones was built, color segmentation was carried out, and a number of color features were calculated, with the help of which it is possible to identify structural components by the color tone of the objects under study. At the same time, the presence of the L and Q components in the color system makes it possible to produce luminance - color segmentation.

Components L, C and Q are stored in corresponding matrices (arrays) **L, C** and **Q.** This increases the speed of access to visual information and allows you to process it separately without increasing the memory footprint. By combining the obtained matrices with the structure describing the wavelet-decomposition of the brightness matrix, a model of visual information was obtained in the analysis and processing system.

The structure of the wavelet level contained four arrays - matrices of wavelet coefficients $\mathbf{W}_{HH}$, $\mathbf{W}_{HL}$, $\mathbf{W}_{LH}$, $\mathbf{W}_{LL}$, fields dw and dh, which determine the size of the matrix coefficients, the number of the decomposition level Level (1, 2,…, n), links of the current level of decomposition. With multilevel decomposition, $\mathbf{W}_{LL}$ are filled only at the last level n; in previous levels, they are subtracted by inverse transformations.

A number of algorithms for processing images of macrostructures have been developed and investigated, which make it possible to automate the processing of visual information of the object under study, performing preliminary processing, segmentation and image recognition.

Testing the algorithm on a series of images showed a decrease in the error of brightness - color segmentation due to a decrease in illumination unevenness by an average of 2 times.

An algorithm for recovering a focused image from a series of visual information based on wavelet analysis is investigated. The algorithm allows to reduce the time of image construction by 2 times in comparison with the well-known algorithm based on the Fourier transform and to reduce the reconstruction error by 4, 5 times.

An algorithm has been developed for identifying defects in the microstructure of the object under study (for example, a scratch, pollution) and control of the structure (for example, the porosity of a product) in the image. All these defects and properties. are outside the focal point of the plane. Therefore, the developed algorithm makes it possible to reduce the time for detecting defects in comparison with the manual method and reduce operator fatigue. Studies have shown that if there are 8 defects in the image, the processing time of visual information is reduced by an average of 5 times.

The questions of construction and research of texture features applicable in the automated control system for classification and recognition tasks are considered [32-34]. To calculate the texture features, the wavelet transform coefficients were used, which made it possible to significantly reduce the effect of noise on their values when calculating these features.

The distance between adjacent brightness values is determined by the level of decomposition. We considered the following features v, calculated from the histogram of the densities P(w) of the distribution of wavelet coefficients of the value w: $\sum P(w)$ log P(w) –entropy, $\sum P^2(w)$ –energy, $\tilde{v}\ \sum wP(w)$ - expected value,

$\sqrt{\sum (w-\bar{v})^2 \; P(w)}$ - CKO, LIM - measure of Lorentz information,
NPIM - standardized measure of information.

Features are calculated at each level of decomposition for each matrix of coefficients W_{HH}, W_{HL}, W_{LH}, W_{LL}. For their invariance to image rotation and to reduce their number, the features calculated using these high-frequency coefficients for each level are averaged. With a three-level decomposition, 24 features are calculated.

Also, the brightness-color, brightness-texture and contour-structural algorithms for the selection of elements in the visual information processing system were investigated, which together allow to process a large variety of classes of images of microstructures of various objects of research and to minimize errors in measuring their parameters (Table 1). Brightness-color segmentation was performed simultaneously using two histograms L and Q. A pixel was considered to be related to an object if its tone Q and brightness L fell within the specified range. Unlike the known ones, this algorithm processes images containing both color and grayscale areas with a lower error, and allows to reduce the segmentation time on average by 5 times due to the use of a data structure representation model.

Table 1

Characteristics of segmentation algorithms

Algorithm	Application	Features
Bright color	For images of structures that do not contain complex structural components	Has the fastest performance among the rest, allows you to select an object by color
Brightness-textural	For research objects containing complex structural components	Eliminates complex segmentation of areas with a homogeneous structure
Contour-structural	For images. containing object outlines	Reduces the error in measuring geometric parameters and the effect of the error

The luminance texture segmentation algorithm includes two processing stages. At the first, "rough" segmentation occurs with recursive division of the image into smaller areas. Subdividing continues until all regions are uniform or the minimum region size is reached. Then areas with a homogeneous texture were combined into clusters by the similarity criterion. As a criterion for the similarity of textures of two regions A and B, the difference was used [35]:

$$J_{AB} = \frac{(N_A+(k_A+1)(l_A+1))v_A+(N_B+(k_B+1)(l_B+1))v_B}{N_A+N_B} - (1+\frac{(k_A+1)(l_A+1)+(k_B+1)(l_B+1)}{N_A+N_B})v_C,$$

where: k, l - size of the area in pixels (the index is determined by the area), kl = N;

v - a feature used in comparison (by area A, B or C);

C —the area obtained by combining areas A and B.

The textures of the regions are recognized as similar if $[J_{AB}] \leq q$, where the q value determines the homogeneity threshold. The difference J_{AB} is based on the principle of comparing the change in a feature before combining the regions v_A and v_B, and after combining v_C.

The number of pixels k, l and sizes N take into account the ratio of these areas and the contribution of each of them to the features of the area C. The area is checked for homogeneity by dividing it into four parts and determining the similarity of the texture characteristics of all four parts. In case of similarity, the area is assumed to be homogeneous.

In the second stage, areas that do not contain textures are combined according to the criterion of the average brightness level. The boundaries of the regions are refined based on the analysis of histograms of the joint occurrence of pixels at the boundaries and within the region. Experimental studies of the algorithm have established its effectiveness for processing images of objects under study, both containing complex structural components, and not containing them.

The contour-structural segmentation algorithm is an expression that describes the areas of a sharp change in brightness between areas and lines. When describing the contours with primitives (a set of primitives contains a "straight", "arc", "point"), the adequacy of this description is assessed by measuring the deviation of the length ε1, and the deviation of the form εf. The length deviation is determined by the difference between the length of the primitive and the length of the contour segment

$$\varepsilon_l = D - \sum_i \sqrt{\Delta x_i^{\,2} + \Delta y_i^{\,2}},$$

where:

Δx_i и Δy_i - the distance between a pair of adjacent pixels;

D - is the length of the primitive.

The shape deviation is defined as the sum of the deviations h of each point of the contour

$$\varepsilon_\Phi = \sum_i h_i$$

The primitive adequately reflects the shape of the contour if the parameters ε 1 and εph do not exceed the specified thresholds. If the primitive does not give an adequate description of the contour, then the segment is further subdivided and inflection points are introduced. Upon completion of the partition, the restoration of the contour breaks in the image is performed. Contour-structural segmentation increases the accuracy of measuring the perimeters of contour objects by 10 times, and their areas by 4 times.

The developed algorithms for automated processing of images of the objects under study produce processing, segmentation and image recognition.

3. Results and discussion

Based on the results of the studies performed, a model of the visual information processing system was developed. From the object under study, a video stream arrives at the input of the model of the visual information processing system, which is first selected as a sequence of frames (frames) of a fixed dimension. Then there is a preliminary processing of the information received from the object, its segmentation, extraction of features and decision-making. Here, the traditional number of homogeneous segments can be applied, into which the object is divided using the principle of maximum posterior probability. The result obtained is then used in the adaptation block. The proposed model of the visual information processing system was implemented in the Qt Creator environment in C ++.

4. Conclusions

The use of the obtained model of the visual information processing system makes it possible to automate the analysis and processing of images of the objects under study and use the results obtained for the further development of an intelligent automated system for quality control of various objects.

References

[1] Balykhin M.G., Blagoveshchenskaya M.M., Blagoveshchenskiy I.G., Makarovskaya Z.V., Nazoykin E.A. (2019). Automation of vacuum freeze-drying products using the combined control method. News of higher educational institutions, 2 (380). Technology of the textile industry *[Tekhnologiya tekstil'noy promyshlennosti]*, 133-137.

[2] Balykhin M.G., Blagoveshchenskiy I.G. (2018). Using the Prewitt method in the development of digital video frame processing algorithms. In the collection The current state and prospects for the development of packaging in the food industry Conference proceedings with international participation. Izdatel'skiy kompleks MGUPP *[Publishing complex MGUPP]*, 76 – 79.

[3] Balykhin M.G., Blagoveshchenskiy I.G. (2018). The use of various methods of formalizing the source images to highlight information in the processing of digital video frames. In the collection The current state and prospects for the development of packaging in the food industry Conference proceedings with international participation. Izdatel'skiy kompleks MGUPP *[Publishing complex MGUPP]*, 79 – 82.

[4] Balykhin M.G., Blagoveshchenskiy I.G. (2018). Using Prewitt's method in the development of algorithms for processing digital video frames. In the collection The current state and prospects for the development of packaging in the food industry Conference proceedings with international participation. Izdatel'skiy kompleks MGUPP *[Publishing complex MGUPP]*, 76 – 79.

[5] Balykhin M.G., Blagoveshchenskiy I.G. (2018). Quality control of confectionery products using various image processing methods. *Materials of the Conference with international participation* (pp. 82 – 85). Publishing complex MGUPP *[Publishing complex MGUPP]*.

[6] Balykhin M.G., Blagoveshchenskiy I.G. (2018). The use of technical vision in automated process control systems for food production with the identification of the main stages of development. *In the collection: Current State and Prospects for the Development of Packaging in the Food Industry Conference Materials with International Participation* (pp. 65 – 68). Publishing Complex MGUPP *[Publishing complex MGUPP]*.

[7] Balykhin M.G., Borzov A.B., Blagoveshchenskiy I.G. (2017). The architecture and basic concept of creating an intelligent expert system of food quality control. Food industry *[Pishchevaya promyshlennost']*, 11, 60 - 63.

[8] Balykhin M.G., Borzov A.B., Blagoveshchenskiy I.G. (2017). Methodological foundations of creating expert systems for monitoring and forecasting the quality of food products using intelligent technologies (pp. 128 – 336). Frantera. [Frantera publishing house].

[9] Blagoveshchenskaya M.M. (2009). Basics of stabilization of the preparation of multicomponent masses. Izdatel'stvo Frantera [Frantera publishing house].

[10] Blagoveshchenskaya M.M., Santon Kunnikhan M.P. (2017). The structure of dosing control systems using neural networks. *In the collection of conference materials: University Student Conference of Students and Young Scientists "Science Day"* (pp. 263 – 267). Izdatel'skiy kompleks MGUPP *[Publishing complex MGUPP]*.

[11] Blagoveshchenskaya M.M., Sulimov V.D., Shkapov P.M. (2010). Methodology for developing the fundamentals of modeling and diagnostics of hydromechanical systems of food production according to their dynamic characteristics // High intellectual technologies and innovations in education and science: Materials of the XVII International Scientific Method. conf. February 11-12, 2010, Izdatel'stvo SPbGPU *[Publishing House SPbSPU], 95-98.*

[12] Blagoveshchenskaya M.M., Zlobin L.A. (2005). *Information technology of control systems of technological processes* (128 - 384). Vysshaya shkola [High school].

[13] Blagoveshchenskaya M.M., Petryakov A.N., Savostin S.D., Blagoveshchenskiy I.G. (2017). The use of object shape recognition algorithms to automate the control of organoleptic quality indicators of feed production. In the book: The Development of the Food and Processing Industry of Russia: Personnel and Science. Izdatel'skiy kompleks MGUPP *[Publishing complex MGUPP]*, 194 – 196.

[14] Blagoveshchensky V.G., Krylova L.A., Maksimov A.S. (2017). Development of a hardware and software complex for monitoring halva production. In the book: Development of the food and processing industry in Russia: personnel and science. Izdatel'skiy kompleks MGUPP [*Publishing complex MGUPP*], 196 -199.

[15] Blagoveshchenskiy I.G. (2015). Using a computer vision system to monitor online the quality of raw materials and finished products of the food industry. Pishchevaya promyshlennost' *[Food industry]*. 6, 9-13.

[16] Blagoveshchenskiy, I.G., (2017). Automated expert control system in the flow of fondant candy quality indicators using neural network technologies and computer vision systems. Izdatel'stvo Frantera [Frantera publishing house].

[17] Blagoveshchenskiy I.G., Sulimov V.D., Shkapov P.M., Blagoveshchenskay M.M. (2018). HYBRID ALGORITHMS FOR OPTIMIZATION AND DIAGNOSTICS OF HYDROMECHANICAL SYSTEMS USED IN FOOD PRODUCTION BIOTECHNOLOGY. *В сборнике: IOP Conference Series: Materials Science and Engineering Сер. "Fundamental and Applied Problems of Mechanics – 2017*, 012039 - 012048.

[18] Blagoveshchenskiy I.G., Makarovskaya Z.V., Blagoveshchenskaya M.M., Chuvakhin S.V., Mitin V.V. (2019). The use of a digital video camera as an intelligent sensor of an automatic control system for the molding of granular food masses. In the collection: Modern problems of engineering and technology of food production. Materials of the XX International Scientific and Practical Conference. Izdatel'skiy kompleks MGUPP *[Publishing complex MGUPP]*, 71-75.

[19] Blagoveshchensky I.G., Nosenko S.M. (2015). Expert intellectual system for monitoring the process of forming fondant sweets using the technical vision system // Pishchevaya promyshlennost' *[Food Industry]*, 6, 32 - 36.

[20] Bychkov I.A., Blagoveshchenskaya M.M., Nosenko A.S., Blagoveshchenskiy I.G. (2015). The method of generalized interval estimates to support group expert decisions in the face of uncertainty. Khraneniye i pererabotka sel'khozsyr'ya *[Storage and processing of agricultural raw materials]*. 4, 63-65.

[21] Fomushkin V.I., Borodin A.V., Blagoveshchensky I.G., Karelina E.B. (2015). Computer-based

prediction of the risks of contamination by populations of microorganisms of meat raw materials during storage. Khraneniye i pererabotka sel'khozsyr'ya *[Storage and processing of agricultural raw materials]*. 7, 12-16.

[22] Garev K.V., Blagoveshchensky I.G., Nazoykin E.A., Blagoveshchensky V.G., Makarovskaya Z.V. (2019). Using technical vision as an innovative solution in smart home systems. In the conference proceedings: Intelligent systems and technologies in the food industry. Izdatel'skiy kompleks MGUPP *[Publishing complex MGUPP]*, 47-52.

[23] Goncharov K.A., Blagoveshchensky I.G., Nazoykin E.A., Blagoveshchensky V.G., Makarovskaya Z.V. (2019). Using the OPENCV library for working with technical vision. In the conference proceedings: Intelligent systems and technologies in the food industry. Izdatel'skiy kompleks MGUPP *[Publishing complex MGUPP]*, 53-60.

[24] Ivanov Y.V., Blagoveshchenskaya M.M., Blagoveshchenskiy I.G. (2012). Automation of the process of molding candy masses using a digital video camera (CVC) sensor. *Materials of the first international scientific and practical conference - the exhibition "Planning and ensuring the training and retraining of personnel for the food industry and medicine."* (pp. 215 – 218). Publishing complex MGUPP *[Publishing complex MGUPP]*.

[25] Karelina E.B., Blagoveshchenskaya M.M., Blagoveshchensky V.G., Klekho D.Yu., Blagoveshchensky I.G. (2019). Integration of adaptive management in the technological processes of the food industry. *In the conference proceedings: Intelligent Systems and Technologies in Food Industries*, 81-89.

[26] Kharitonova P.N., Karelina E.B., Blagoveshchenskiy V.G., Klekho D.Yu., Blagoveshchenskiy I.G. (2019). The introduction of digital double control in technological production. In the conference proceedings: Intelligent systems and technologies in the food industry. Izdatel'skiy kompleks MGUPP *[Publishing complex MGUPP]*, 171-180.

[27] Krylova L.A., Blagoveshchenskiy V.G., Tatarinov A.V. (2017). Development of intelligent hardware and software systems for monitoring the separation of dispersed food masses based on intelligent technologies. *In the book: The Development of the Food and Processing Industry of Russia: Personnel and Science* (pp. 199 – 201). Publishing complex MGUPP *[Publishing complex MGUPP]*.

[28] Nazoykin E.A., Blagoveshchenskiy I.G. (2019). The use of simulation methods to identify the processes of the dough preparation department at the bakery. In the collection: Ninth All-Russian Scientific and Practical Conference on Simulation and its Application in Science and Industry. Conference proceedings. Izdatel'skiy kompleks MGUPP *[Publishing complex MGUPP]*, 468-472.

[29] Nazoykin E.A., Blagoveshchensky I.G., Sincha V.M., Zhirov M.V., Mitin V.V. (2019). The use of simulation to identify the state of enterprises in the food industry. In the conference proceedings: Intelligent systems and technologies in the food industry. Izdatel'skiy kompleks MGUPP *[Publishing complex MGUPP]*, 147-155.

[30] Petrov A.Yu., Blagoveshchenskaya M.M., Blagoveshchenskiy V.G., Ionov A.V., Blagoveshchenskiy I.G. (2019). The main principles in building a computer vision system in the baking industry. In the conference proceedings: Intelligent systems and technologies in the food industry. Izdatel'skiy kompleks MGUPP *[Publishing complex MGUPP]*, 121-126.

[31] Petrov A.Yu., Blagoveshchenskaya M.M., Makarovskaya Z.V., Mitin V.V., Blagoveshchenskiy I.G. (2019). The use of neural networks in the regulation of additives in bakery products. In the conference proceedings: Intelligent systems and technologies in the food industry. Izdatel'skiy kompleks MGUPP *[Publishing complex MGUPP]*, 172-178.

[32] Petryakov A.N., Blagoveshchenskaya M.M., Blagoveshchensky V.G., Krylova L.A. (2018). The use of object-oriented programming methods to control the quality indicators of confectionery products. *Zhurnal "Konditerskiye izdeliya i khlebopecheniye" [Confectionery and Bakery Magazine]*, no. 5-6 (176), 21-23.

[33] Petryakov A.N., Blagoveshchenskaya M.M., Blagoveshchenskiy V.G., Mitin V.V., Blagoveshchenskiy I.G. (2019). Improving the quality of identification and positioning of an object in digital stereo images using algorithms for constructing a depth map. In the conference proceedings: Intelligent systems and technologies in the food industry. Izdatel'skiy kompleks MGUPP *[Publishing complex MGUPP]*, 133-138.

[34] Petryakov A.N., Blagoveshchenskaya M.M., Nosenko A.S., Blagoveshchenskiy I.G. The use of genetic algorithms and particle swarm methods in the production of animal feed. Pishchevaya

promyshlennost' *[Food industry]*. 6, 18-21.
[35] Savostin S.D., Blagoveshchenskaya M.M., Blagoveshchensky I.G. (2016). Automation of control of flour quality indicators during grinding using intelligent technologies. Izdatel'stvo Frantera *[Frantera publishing house]*.

IOP Conf. Series: Materials Science and Engineering 1191 (2021) 012013 doi:10.1088/1757-899X/1191/1/012013

Vibrations of rigid cantilevered plates in a rectangular tank filled with liquid

Goncharov D.A.[1], **Pozhalostin A.A.**[1]

[1]Bauman Moscow State Technical University, 2-ya Baumanskaya str. 5, b.1, Moscow 105005, Russia
E-mail: goncharov@bmstu.ru

Abstract. The article presents the developed analytical (exact) methods for solving problems of calculating the frequencies of natural vibrations of rigid and elastic partitions in a rigid rectangular tank filled with liquid. The results of solving these problems can be used in the design of hydraulic structures, as well as in shipbuilding when analysing the dynamics of structural elements of large oil tankers and gas carriers.

1. Introduction

4 cases were considered. In the first case, the tank is closed at one end by an elastically sealed rigid plate. In the second, the plate is elastically fixed in the middle of the tank. In the third case, two plates from cases 1 and 2 are located in the tank. In the fourth case, the solution of the boundary value problem for an elastic plate-console located in the middle of the tank is considered.

In the latter case, a frequency equation is obtained, the left side of which is a meromorphic function with an infinite number of simple poles

The article presents an analytical solution to the problem of small oscillations of a liquid in a rigid tank with an elastic baffle. This mechanical system is a model for various oil tankers. The frequency transcendental equation of the problem is obtained in closed form under certain assumptions, the left side of which is a meromorphic function with an infinite number of simple poles. This result is achieved by expanding the homogeneous solution into a series in terms of the eigenfunctions of the Laplace equation in a rectangular coordinate system. One of the first specialists to obtain results in this problem of hydroelasticity was P.K. Ishkov. [4].

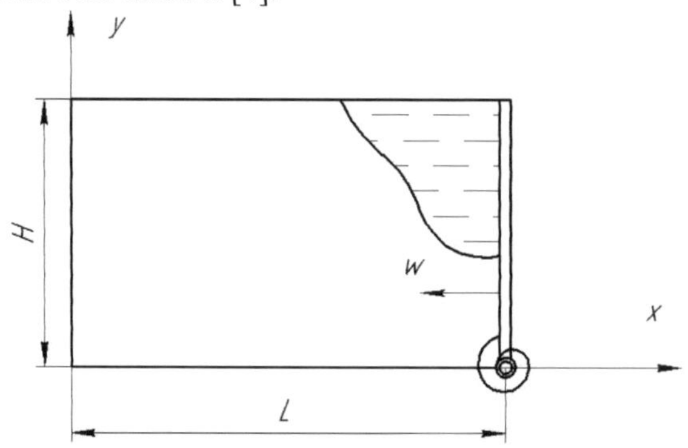

Figure 1. Schematic representation of the tank

To obtain analytical solutions, we used the method of expanding the solution of a homogeneous equation in a series in the complete system of eigenfunctions of the Laplace equation.

Later, this technique was successfully used by L.S. Leibenzon. [3], Balabukh L.I. [2] and other researchers.

In the first three cases, the solution is constructed in such a way that the associated moments of inertia from the fluid.

In the latter case, a rigorous solution of the boundary value problem is obtained within the accepted assumptions, namely, for the plane problem and small fluctuations.

2. Boundary value problem

We consider the plane problem of vibrations of a tank closed by an elastically sealed partition is considered. The partition is absolutely rigid, weighing m_0 performs angular oscillations according to the law $\varphi(t)$. We put $w(t) = y\varphi(t)$. There is an elastic connection, torsional stiffness C0. The liquid fills a rectangular tank with a length L to a height H.

This boundary value problem is solved under the following assumptions: the fluid is assumed to be ideal and incompressible, its motion is potential, with the velocity potential Φ. The dynamic pressure of the fluid through the Lagrange-Cauchy integral is determined by the formula $P = \rho\dfrac{\partial \Phi}{\partial t}$.

Here ρ is the density of the liquid. The velocity of a liquid particle will be determined by the relation:

$$\upsilon = -grad\Phi \;.$$

The velocity potentials are solutions of the Laplace equation (Fig. 1)

$$\frac{\partial^2 \Phi_i}{\partial x_i^2} + \frac{\partial^2 \Phi_i}{\partial y_i^2} = 0 \qquad (1)$$

The boundary conditions for the velocity potential are:

$$\left.\frac{\partial \Phi}{\partial x}\right|_{x=0} = 0, \quad \left.\frac{\partial \Phi}{\partial x}\right|_{x=L} = \dot{w}\,. \qquad (2)$$

We will seek the velocity potential in the form:

$$\Phi = \sum_{i=1}^{\infty} X_i(x) Y_i(y)\dot{s}(t)\,. \qquad (3)$$

On the free surface, we take the linearized boundary condition $\Phi|_{y=H} = 0$. Satisfying the boundary conditions (2) and the condition on the free surface, we obtain the velocity potential in the form:

$$\Phi = \sum_{i=1}^{\infty} C_i cosh\mu_i x cos\mu_i y s(t)\,. \qquad (4)$$

With this choice of potential, we focus on plate vibrations with a liquid, while calculating the attached moment of inertia from the liquid. $\mu_i = \dfrac{\pi(2i-1)}{2H}$ mu are the eigenvalues of the considered boundary value problem. The system of eigenfunctions $Y_i(y)$ is a complete, orthogonal system of functions. Since $y\varphi = w$, we expand y in a series in the system of functions $cos\ \mu_i y$,

$y = \sum_{i=1}^{\infty} \alpha_i cos\ \mu_i y$ we get

$$\alpha_i = (-1)^{i+1} H/\mu_i - 1/\mu_i^2 \qquad (5)$$

Taking into account (5), the velocity potential takes the form

$$\Phi = \sum_{i=1}^{\infty} cosh\mu_i x(cos\mu_i y)\left(\frac{\alpha_i\dot{\varphi}}{\mu_i}\right) \tag{6}$$

The differential equation of the rotational motion of the plate under the action of the hydrodynamic pressure of the liquid has the form:

$$J_0\ddot{\varphi} + C\varphi = -\int_0^H \rho\sum_{i=1}^{\infty} \mu_i cosh\mu_i L(cos\mu_i y)ydy\ddot{\varphi}.$$

Finally, $J = J_0 + \sum_{i=1}^{\infty}\gamma_i$ is a moment of inertia of the plate taking into account the liquid. The expression for the constant γ is not shown, the moment of forces is also not taken into account.

In the second case, an elastically fixed plate is located in the middle of a rectangular tank of length $L = l_1 + l_2$. This boundary value problem is solved under the same assumptions that were made in the first case. Velocity potential in the left compartment is Φ_1 and in the right is Φ_2 They satisfy the Laplace equation and have the form:

$$\Phi_1 = \sum_{i=1}^{\infty} A_{1i} cos\mu_i y cosh\mu_i x\dot{s}_1, \quad \mu_i = \frac{\pi(2i-1)}{2H}, \tag{7}$$

$$\Phi_2 = \sum_{i=1}^{\infty} B_{1i} cos\mu_i y(D_{1i} cosh\mu_i x + D_{2i} sinh\mu_i x)\dot{s}_2. \tag{8}$$

The boundary condition for the potentials is the equality of the fluid velocities at $x = l_1$.

$$\left.\frac{\partial\Phi_1}{\partial x}\right|_{x=l_1} = \left.\frac{\partial\Phi_2}{\partial x}\right|_{x=l_1}. \tag{9}$$

whence $\dot{s}_1 = \dot{s}_2$. Taking into account (9), the velocity potential functions have the form

$$\Phi_1 = \sum_{i=1}^{\infty} D_{1i}\gamma_i cos\mu_i y cosh\mu_i x\dot{s}_1,$$

$$\Phi_2 = \sum_{i=1}^{\infty} D_{1i} \cos\mu_i y\left(\cosh\mu_i x - \tanh\mu_i L \sinh\mu_i x\right)\dot{s}_1.$$

In this case, the differential equation of plate vibrations will be:

$J\ddot{\varphi} + C\varphi = 0$, where $J = J_0 - \beta_1 + \beta_2$. The constants β are infinite rapidly converging series. They are determined from the formulas

$$\int_0^H p_1 ydy = \beta_1\ddot{\varphi} \text{ and } -\int_0^H p_2 ydy = \beta_2\ddot{\varphi}.$$

In the third case, in a rigid rectangular tank of length $L = l_1 + l_2$ are two rigid rectangular plates elastically fixed as in the first case - one plate is located in the middle of the tank, and the second is elastically fixed at the end of the tank at $x = L$ (Fig. 1). The assumptions under which this boundary value problem is solved are similar to those that were earlier when solving the first two problems. Taking into account the boundary conditions on the free surface, on the bottom and on the walls of the vessel, we obtain the following expressions for the velocity potentials

$$\Phi_1 = \sum_{i=1}^{\infty} A_{1i}\gamma_i \cos\mu_i y \cosh\mu_i x\dot{s}_1,$$

$$\Phi_2 = \sum_{i=1}^{\infty} \cos \mu_i y \left(A_{2i} \cosh \mu_i x + A_{3i} \sinh \mu_i x \right) \dot{s}_1 .$$

Since the velocity of the point on the surface of the middle plate is equal to the velocity of the liquid particle, and the velocity of the point of the side plate 2 is equal to the velocity of the liquid particle of the second potential and the velocities of both potentials are determined in accordance with (9), then after simple transformations and using the expansion y in a series in the system of functions $cos\ \mu_i y$, we get

$$A_{1i}\dot{s}_1 = \beta_{11}\dot{\varphi}_1 + \beta_{12}\dot{\varphi}_2 ,$$
$$A_{2i}\dot{s}_2 = \beta_{21}\dot{\varphi}_1 + \beta_{22}\dot{\varphi}_2 .$$

Here β are some constants whose form is not given.

The differential equations for the vibrations of the plates are as follows:

$$J_1\ddot{\varphi}_1 + C_1\varphi_1 = \gamma_{11}\ddot{\varphi}_1 + \gamma_{12}\ddot{\varphi}_2 ,$$
$$J_2\ddot{\varphi}_2 + C_2\varphi_2 = \gamma_{21}\ddot{\varphi}_1 + \gamma_{22}\ddot{\varphi}_2 .$$

Here γ_{ij} are some coefficients that are rapidly converging series. The equations represent a system of two differential equations as in the case of oscillations of a mechanical system with two degrees of freedom with inertial constraint.

3. Conclusion

The compartments of a rectangular rigid tank are separated by a uniform ideally elastic cantilever-sealed beam of linear mass μ_0 of height H. The material of the beam obeys Hooke's law, its bending is considered to be straight, and EJ_0 is its linear bending stiffness [5], $\Phi(x, y, t)$ is the velocity potential. We will use the Fourier method [4], then the solution to equation (1) has the form similar to examples 1-3, where the constants A_j, E_j, B_j, D_j are to be determined, $s(t)$ is a time function. The differential equation of bending vibrations of the beam will have the form

$$EJ_0 w_x^{IV} - \mu_0\omega^2 w = \beta_1\dot{\Phi}_1 - \beta_2\dot{\Phi}_2 \qquad (10)$$

The velocity potential functions satisfy the boundary conditions on the free surface and the impermeability conditions. Using the property of orthogonality of the vibration modes Φi, we obtain the dependences for arbitrary constants A_j, E_j, B_j, D_j through the constants of the homogeneous solution of the differential equation (10). Satisfying the conditions for fixing the cantilever beam We obtain the frequency equation of the problem (in closed form) whose left side is a meromorphic function.

References

[1] Kochin N.E. and others. Theoretical hydromechanics. M., FN. GIFML. 1963.
[2] Balabukh L.I., Molchanov A.G.. *PMT*, **Vol. 6**, 1966, p. 1098-1102.
[3] Leibenzon L.S. *Reports of the USSR Academy of Sciences*, 1951, **Vol. 1**, p. 125-133.
[4] Ishkov P.K. PMM. 1937, **Vol. 1**, pp. 1 - 18
[5] Strelkov S.P. Introduction to the theory of oscillations, Nauka, 1964, 437 p.

Solution of two-dimensional problems of nonlocal elasticity theory by the finite element method

Kuvyrkin G.N.[1], **Savelyeva I.Yu.**[1], **Sokolov A.A.**[1]

[1]Bauman Moscow State Technical University, 2-nd Baumanskaya st, 5-1, 105005, Russian Federation

E-mail: fn2@bmstu.ru

Abstract. In this work, based on a model of the nonlocal theory of elasticity, the solution of a two-dimensional problem of a plane stressed rectangular region is considered. The use of approaches of generalized mechanics of a continuous medium made it possible to consider the effects of nonlocality of the medium in space and to obtain integrodifferential equations as constitutive relations, to which well-studied numerical methods are applicable, in particular the finite element method.

1. Introduction

Modern mechanical engineering widely uses new materials that are capable of withstanding various high-intensity external influences. Such materials have unique physical, mechanical and thermophysical properties. These materials include a variety of composites, consolidated structurally sensitive materials, which, for example, are obtained by compaction of nanopowders, deposition on a substrate, crystallization of amorphous alloys, and some other methods. [1, 2].

An important stage in the creation of such materials is the construction of mathematical models that allow describing their behavior under various external influences since when considering micro- and nanostructured materials, classical approaches of continuum mechanics cease to work. This is due to the fact that in bodies of this size, effects begin to prevail, which do not have a significant effect in bodies of large sizes [3, 4].

There are many approaches to building models for nanostructured materials: quantum-mechanical approaches, molecular dynamics methods and Monte Carlo models [4]; approaches of generalized continuum mechanics, including the theory of effective properties [5 - 7], mixture models [8, 9], models based on averaging procedures [10], Cosserat-type micropolar media [11] and statistical models [12]; hybrid molecular mechanical models combining molecular dynamics methods and macroscale models [13, 14].

2. Statement of the problem and basic relations

In an arbitrary domain $S \subset \mathbb{R}^2$ with a piecewise smooth boundary ∂S, the equilibrium equation for a continuous medium has the form [15]

$$\nabla \cdot \hat{\boldsymbol{\sigma}} + \boldsymbol{b} = \boldsymbol{0}, \qquad (1)$$

where $\nabla = \partial / \partial x_i \boldsymbol{e}_i$, $i = \overline{1,2}$ is differential operator nabla; $\boldsymbol{e}_i$, $i = \overline{1,2}$ are unit vectors of the coordinate axes; x_i, $i = \overline{1,2}$ are vector components $\boldsymbol{x}$; $\boldsymbol{b} = b_i \boldsymbol{e}_i$, $i = \overline{1,2}$ is body force vector; $\hat{\boldsymbol{\sigma}} = \sigma_{ij} \boldsymbol{e}_i \otimes \boldsymbol{e}_j$, $i, j = \overline{1,2}$ is stress tensor, which is defined as follows

$$\hat{\boldsymbol{\sigma}} = p_1 \hat{\boldsymbol{C}} \cdot\cdot \hat{\boldsymbol{\varepsilon}} + p_2 \iint\limits_{S'(\boldsymbol{x}) \cap S} \varphi(|\boldsymbol{x} - \boldsymbol{x}'|) \hat{\boldsymbol{C}} \cdot\cdot \hat{\boldsymbol{\varepsilon}} dS'(\boldsymbol{x}). \tag{2}$$

Here $\hat{\boldsymbol{\varepsilon}} = \varepsilon_{ij} \boldsymbol{e}_i \otimes \boldsymbol{e}_j$, $i, j = \overline{1,2}$ is elasticity tensor; $\hat{\boldsymbol{C}} = C_{ijkl} \boldsymbol{e}_i \otimes \boldsymbol{e}_j \otimes \boldsymbol{e}_k \otimes \boldsymbol{e}_l$, $i, j, k, l = \overline{1,2}$ is elastic coefficient tensor; $p_1 > 0$ and $p_2 \geq 0$ are weight fractions such that $p_1 + p_2 = 1$; φ is nonlocal influence function, some normalized positive function in the domain $S'(\boldsymbol{x})$; $S'(\boldsymbol{x})$ is the area of non-local influence; $\boldsymbol{x}' \in S'(\boldsymbol{x})$.

We will consider the boundary conditions of the first and second kind, also called kinematic and force, respectively [14]

$$\boldsymbol{u}|_{\Gamma_1} = \boldsymbol{d}(\boldsymbol{x}), \qquad \hat{\boldsymbol{\sigma}} \cdot \boldsymbol{n}|_{\Gamma_2} = \boldsymbol{p}(\boldsymbol{x}), \tag{3}$$

where $\Gamma_1, \Gamma_2 \subset \partial S$; $\Gamma_1 \cup \Gamma_2 = \varnothing$; $\boldsymbol{u} = u_i \boldsymbol{e}_i$; $i = \overline{1,2}$ is displacement vector; $\boldsymbol{d}(\boldsymbol{x}) = d_i(\boldsymbol{x}) \boldsymbol{e}_i$ and $\boldsymbol{p}(\boldsymbol{x}) = p_i(\boldsymbol{x}) \boldsymbol{e}_i$, $i = \overline{1,2}$ are some functions that set displacements and pressures at the boundaries, respectively.

Introduce the finite element model mesh S_h on the area S. Then, applying the standard finite element approach [16, 17] to equation (1), after substituting (2) into it and passing to the index notation, we obtain the following equality [18]

$$\begin{gathered} p_1 \iint\limits_S \frac{\partial N_n^{(e)}}{\partial x_i} C_{ijkl} \varepsilon_{kl} dS + p_2 \iint\limits_S \frac{\partial N_n^{(e)}}{\partial x_i} \iint\limits_{S'(\boldsymbol{x}) \cap S} \varphi(|\boldsymbol{x} - \boldsymbol{x}'|) C_{ijkl} \varepsilon_{kl} dS'(\boldsymbol{x}) dS - \\ -\oint\limits_{\Gamma_2} N_n^{(e)} p_i d\Gamma + \iint\limits_S N_n^{(e)} b_i dS = 0, \quad i, j, k, l = \overline{1,2}, \quad n = \overline{1, M^{(e)}}, \quad (e) \in S_h, \end{gathered} \tag{4}$$

where $\{N_i^{(e)}\}_{i=1}^{M^{(e)}}$ is set of element basis functions (e).

The Cauchy relations determine the components of the elastic strain tensor $\hat{\boldsymbol{\varepsilon}}$ [14]

$$\varepsilon_{ij} = \frac{1}{2} \left(\frac{\partial u_i}{\partial x_j} + \frac{\partial u_j}{\partial x_i} \right), \quad i, j = \overline{1,2}. \tag{5}$$

Then we write the approximation of the displacements in the following form
$\boldsymbol{u}(\boldsymbol{x}) \approx N_i^{(e)} \boldsymbol{u}_i^{(e)}, i = \overline{1, M^{(e)}}$, where $\boldsymbol{u}_i^{(e)}$ are desired displacements in element nodes (e).

In the case of a linear isotropic body, the elastic tensor components $\hat{C}$ are determined as follows [15]

$$C_{ijkl} = \lambda \delta_{ij} \delta_{kl} + \mu \left(\delta_{ik} \delta_{jl} + \delta_{il} \delta_{jk} \right), \quad i, j, k, l = \overline{1,2}, \tag{6}$$

where λ and μ are Lamé constants, which can be determined through Poisson's ratio ν and Young's modulus E as follows

$$\lambda = \frac{\nu E}{(1+\nu)(1-2\upsilon)}, \quad \mu = \frac{E}{2(1+\nu)}. \tag{7}$$

3. Choice of the nonlocal influence function

This formulation of the problem (2) gives a very wide scope for choosing the zone of non-local influence $S'(\boldsymbol{x})$ and the corresponding nonlocal influence function φ. In the case of an isotropic body, it is reasonable to assume that the function of non-local influence should be symmetric along each of its axes, therefore, the region of non-local influence should be chosen in the form of a circle. Usually, the Gaussian normal distribution function is chosen, and the zone $S'(\boldsymbol{x})$ is approximated by the three-sigma rule, but the calculation of exponentials is very expensive in terms of computational resources, therefore, in this work, we will consider relatively "cheap" polynomial analogs with a given radius of influence r. A family of functions of the following form will be considered as polynomial functions of nonlocal influence

$$\varphi_{p,q}(|\boldsymbol{x}-\boldsymbol{x}'|)=\begin{cases}\dfrac{p}{2\pi r^2 B(2/p,q+1)}\left(1-\left(\dfrac{|\boldsymbol{x}-\boldsymbol{x}'|}{r}\right)^p\right)^q, & |\boldsymbol{x}-\boldsymbol{x}'|\le r,\\ 0, & |\boldsymbol{x}-\boldsymbol{x}'|>r,\end{cases} \tag{8}$$

where $p,q>0$; $r>0$ is the radius of non-local influence; B is Euler's beta function.

By varying the parameters p and q, we can notice that with an increase in the parameter p the distribution of the influence function tends to the constant $\left(\pi r^2\right)^{-1}$, while as the parameter increases q the distribution tends to the Dirac delta function, which, when substituted in (2), will give us the classical Hooke's law. Graphs of the influence function in the section along the axis of symmetry at various parameters p and q are shown in Fig. 1 and 2.

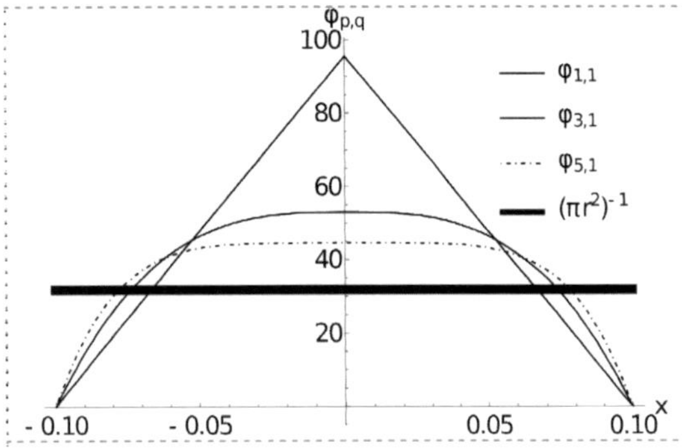

Fig. 1. Graphs of the influence function at various parameters p.

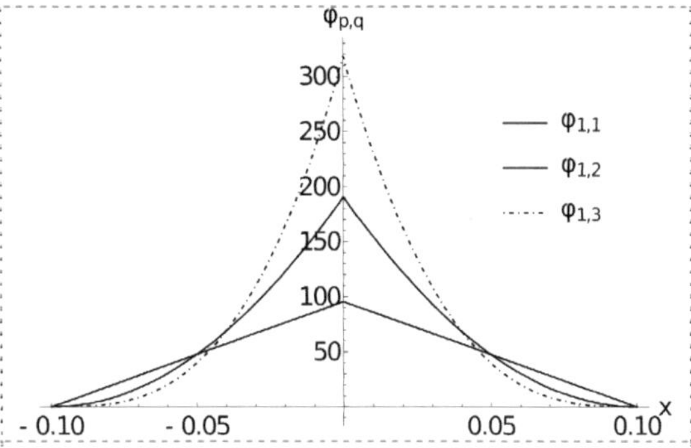

Fig. 2. Graphs of the influence function at various parameters q.

4. Numerical results

Further, in the calculations, we will use the following dimensionless parameters and variables

$$\bar{\sigma}_{11}=\frac{\sigma_{11}}{\sigma},\quad \bar{x}_1=\frac{x_1}{L},\quad \bar{u}_1=\frac{E}{\left(1-\nu^2\right)p}\frac{u_1}{L}. \tag{9}$$

Here L is characteristic area size; σ is normalizing factor for stresses; the bar above the variables means they are dimensionless.

As a test problem, on which we will study the influence of the main parameters of the model, we choose the Neumann problem. For domain $S=[0,1]\times[0,1]$, with a uniform mesh introduced on it S_h, consisting of 10000 quadratic serendipity elements, the following boundary conditions are set

$$\bar{\boldsymbol{\sigma}} \cdot \boldsymbol{n}\,|_{\bar{x}_1=0} = -1, \quad \bar{\boldsymbol{\sigma}} \cdot \boldsymbol{n}\,|_{\bar{x}_1=1} = 1, \quad \bar{\boldsymbol{\sigma}} \cdot \boldsymbol{n}\,|_{\bar{x}_2=0,1} = 0. \tag{10}$$

Naturally, the problem in this formulation has infinitely many solutions, so two additional geometric conditions are required. They are chosen so that the solution acquires symmetry about the center of the region and is fixed along its main axes

$$\bar{u}_i\,|_{\bar{x}_i=0.5} = 0, \quad i = \overline{1,2}. \tag{11}$$

In the classical formulation, due to the symmetry of the problem, it would be possible to dispense with the consideration of only a quarter of the region, but in the nonlocal formulation, one has to perform the procedure for finding the nearest neighbors, which is necessary to approximate the region $S'(\bar{\boldsymbol{x}})$, and without additional modifications, it is not allowed to consider part of the area S. The material parameters are set as follows: Poisson's ratio ν takes equal 0.3, dimensionless module $\bar{E} = 21$.

The classical solution in setting (10) gives us linear solutions for displacements and constants for deformations and stresses, but when nonlocal effects are taken into account, graphs become more complicated. When non-local effects are taken into account, the so-called "edge effect" arises, which is the stronger, the greater the contribution of the non-local influence. The results of calculations for different weight parameters can be seen in Fig. 3 and 4. The design parameters, as well as the selected sections are shown in figures.

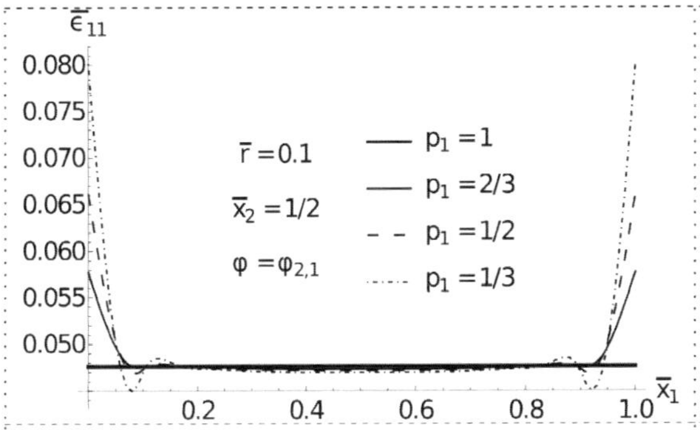

Fig. 3. Strain with a variety of weights.

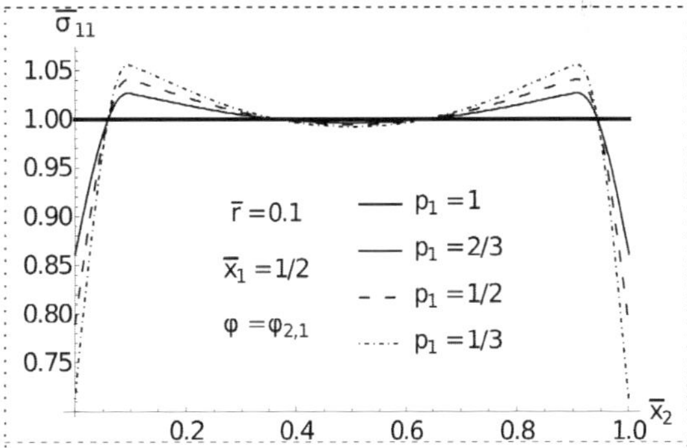

Fig. 4. Stress with the variety of weights.

The emerging edge effect explains that when approaching the boundary ∂S zone of non-local influence $S'(\bar{\boldsymbol{x}})$ is cut off due to which the fraction of the integral term decreases, which in turn leads to a greater influence of the classical Hooke's law. It is worth noting that in the center of the region, that is, in places that are far from the border at a distance greater than the radius of influence $\bar{r}$, the solution taking into account nonlocal effects does not differ much from the classical. That is, as the radius increases, the edge effect also becomes wider. This statement can be confirmed by trying to vary the radius of nonlocal influence $\bar{r}$. The calculation results are shown in Fig. 5 and 6.

The study of the family of influence functions (8) shows that the deviation from the classical solution becomes the stronger, the larger the parameter p. Increase parameter q on the contrary reduces the deviation. This fact is illustrated in Fig. 7 and 8, where, in turn, not the deviations themselves are shown, but the relative difference of solutions. The cross section with the largest deviation was also chosen to show that taking into account non-local effects can significantly change the solution.

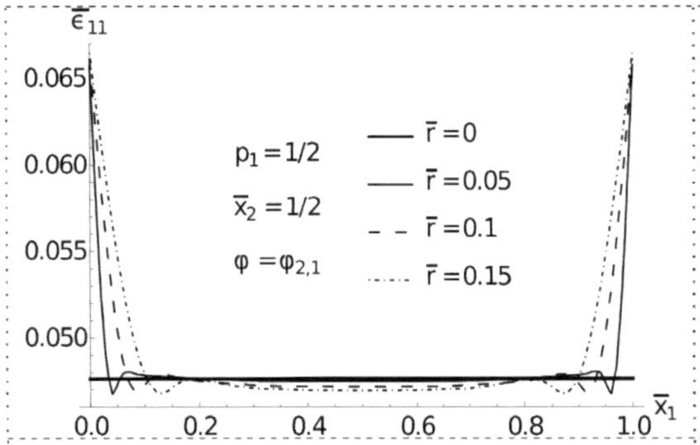

Fig. 5. Strain with the variation of the radius of influence.

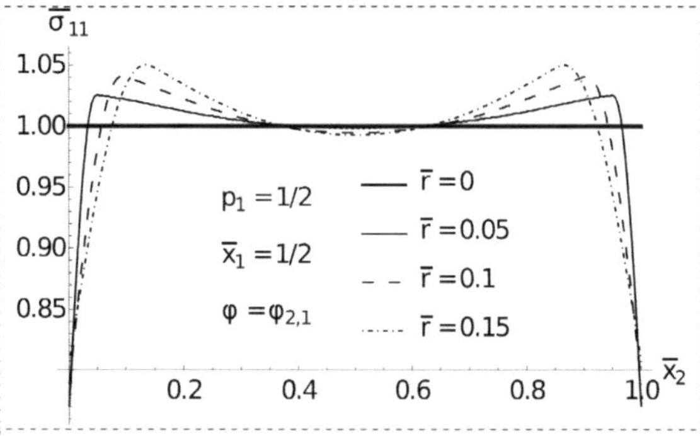

Fig. 6. Stress with the variation of the radius of influence.

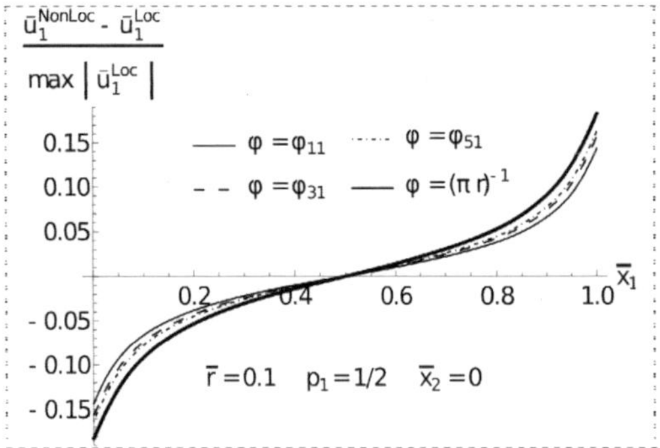

Fig. 7. Relative difference between non-local and local solutions with variation of the parameter p.

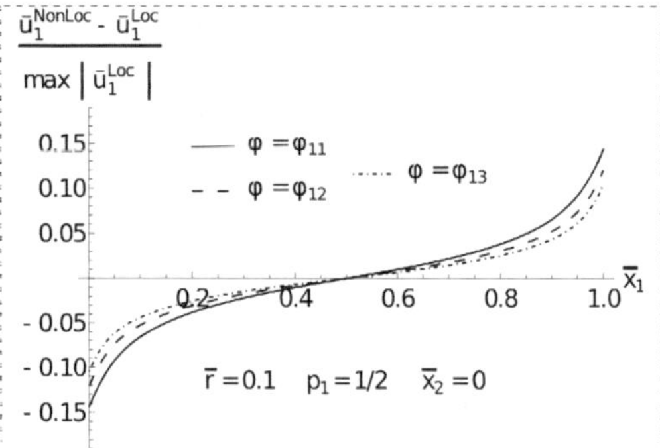

Fig. 8. Relative difference between non-local and local solutions with variation of the parameter q.

5. Conclusion

Using the example of the Neumann problem, the influence of the main parameters of the model was investigated. Influence of weight parameters, the radius of non-local influence, as well as comparison of solutions when choosing different functions of non-local influence. It was shown that taking into account nonlocal effects significantly affects the final results of the solution, especially at the boundary of the region. The article demonstrated only part of the results obtained, displacements u_2 were not shown and other components of the elastic strain tensor and stress tensor due to the fact that for this problem they were not of particular interest, although they also change and have a similar character of changes.

The implementation of the finite element algorithm was written in the C ++ programming language using the OpenMP parallel programming technology. The Eigen linear algebra library was used to store symmetric sparse matrices. The library Intel MKL Pardiso was used as a solver for systems of linear algebraic equations.

Acknowledgments

This work was supported by the Ministry of Science and Higher Education (project № 0705-2020-0032)

References

[1] Thangadurai T.D., Manjubaashini N., Thomas S., Maria H.J. Nanostructured Materials. Springer International Publishing, 2020. 210 p.
[2] Alymov M.I. Zelensky V.A. Production methods and physicomechanical properties of bulk nanocrystalline materials – Elise, 2005 [in Russian].
[3] Gusev A.I. Nanomaterials, nanostructures, nanotechnologiy. Moscow: Fizmatlit, 2005. 416 p [in Russian].
[4] Gopalakrishnan S., Narendar S. Wave Propagation in Nanostructures. Nonlocal Continuum Mechanics Formulations. New York: Springer International Publishing, 2013, 365 p.
[5] Eringen A.C. Nonlocal continuum field teories. New York-Berlin-Heidelberg: Springer-Verlag, 2002, 393 p.
[6] Savelyeva I.Y. Numerical simulation of a thermal shock in an elastic body considering non-locality effects in the medium (2020) Herald of the Bauman Moscow State Technical University, Series Natural Sciences, (90), pp. 20-29.
[7] Zarubin, V.S., Kuvyrkin, G.N., Savel'eva, I.Yu. Mathematical model of a nonlocal medium with internal state parameters (2013) Journal of Engineering Physics and Thermophysics, 86 (4), pp. 820-826.
[8] Ruschitsky Y.Y. Elements of mixture theory. Kiev: Nauka. Dumka, 1991, 160 p [in Russian].
[9] Golovin N.N., Zarubin V.S., Kuvyrkin G.N. Mixed models of composite mechanics. Ch. 1 Thermomchanics and thermoelasticity of a multi-composite mixture. Bulletin of MSTU N.E. Bauman. Ser. Natural Sciences. 2009, №3(34), pp. 36-49 [in Russian].
[10] Bahvalov N.S., Panasenko G.P. Averaging processes in periodic media: mathematical problems in the mechanics of composite materials. Moscow: Nauka, 1984, 352 p [in Russian].
[11] Shermergor T.D. Theory of elasticity of microinhomogeneous media. Moscow: Nauka, 1977. 400 p [in Russian].
[12] Karlicic D., Murmu T., Adhikari S., McCarthy M. Nonlocal Structural Mechanics, 2016, John Wiley and Son Inc., Hoboken, 374 p.
[13] Kunin I. A. Theory of elastic media with a microstructure. Nonlocal theory of elasticity – The science, 1975. – 416 p. [in Russian].
[14] Onami M. Introduction to micromechanics. Per. with japan. / M. Onami, S. Iwashimizu, K. Hank, K. Shiozawa, K. Tanaka – Metallurgy, 1987. – 280 p. [in Russian].
[15] Zarubin, V. S. Mathematical Models of Continuum Mechanics and Electrodynamics / V. S. Zarubin, G. N. Kuvyrkin – IMGTU im. Baumana, 2008. – 512 p. [in Russian].
[16] Zienkiewicz O. The Finite Element Method: Its Basis and Fundamentals. Seventh edition. / O. Zienkiewicz, R. Taylor, J.Z. Zhu – 2013. – 756 p.
[17] Bathe K-J. Finite Element Procedures. Second edition. – 2014. – 1065 p.
[18] Pisano A.A., Sofi A., Fuschi P., 2009. Nonlocal integral elasticity: 2D finite element based solutions. Int. J. Solids Struct. 46, 3836–3849.

Application of the problem of analytical regulation of optimal expansion of drugs in a biological system in an open area

K Z Kaspirovich

Peoples' Friendship University of Russia (RUDN University), Moscow, Russia
The work was supported by RFBR grant № 19-08-00261 A.

Kamilaska95@gmail.com

Abstract. In this paper, by using the parametrization method, i.e., describing the selected elements of the system and elementary effects on it using parameters (parameters can be either continuous or discrete) and establishing analytical and graphical methods of various kinds of dependencies between the entered cellular parameters, a mathematical model of a biological cell is compiled. The construction of the proposed system leads to the solution of a mathematical problem for which the concept of optimality is reduced to achieving the shortest time of arrival in a given state.

1. Introduction

Biological systems are living objects of varying complexity (cells and tissues, organs, organ systems, organisms, and the biosphere as a whole).

As complex, highly organized systems, biological systems are capable of adapting to the external environment, developing, reproducing themselves, and evolving. Since many processes are constantly occurring in any biological system, they are referred to as dynamic systems. Since biological systems can exist only if there is a continuous exchange of energy, matter, and information both between the components of the system and with the environment, they are referred to as open systems. In fact, living systems are in dynamic equilibrium with the environment, and a constant flow of energy from outside is needed to prevent complete equilibrium with the surrounding world.

Any organism can be considered as a complex physical and chemical biological system that exists in the environment in a stationary state. It is this ability of living systems that is in a state of equilibrium with the external environment in continuously changing conditions and determines their survival. In order to maintain such a state, the numerous organs and tissues that make up it must function with such intensity as to ensure its adequate residence in the environment.

The regulatory mechanisms found in living organisms are similar in many ways to the regulatory devices in non-living systems, such as machines. In both cases, stability is achieved through a certain form of control. The application of control theory to biological processes has allowed a deeper understanding of the functional interactions between the components of many physiological mechanisms.

Such science deals, in particular, with the general regularities of regulation in living and inanimate systems. Plant and animal physiologists often use precise mathematical models of control theory to explain the mechanisms of action of biological regulatory systems. The measure of the effectiveness of any control system is the degree of deviation of the regulated parameter from the proper (optimal) level and the speed of return to this level. To implement feedback, it is necessary that the result of the operation of this system is compared with the set value, which is the optimal value of the controlled parameter (variable), and in case of deviation from it, it is changed accordingly. When the system is out of balance, a number of consequences arise that lead to the elimination of this violation and to the return of the system to its original state.

In this article, thanks to the use of the parametrization method, i.e., the description of the selected elements of the system and the elementary effects on it using parameters (parameters can be both continuous and discrete) and the establishment of analytical and graphical methods of various kinds of dependencies between the introduced cellular parameters, a mathematical model of a biological cell is compiled. The construction of the proposed system leads to the solution of a mathematical problem, for which the concept of optimality is reduced to achieving the shortest time of arrival in a given state.

Our goal is to find a control that meets the accepted optimality criterion and guarantees the existence of the specified properties of the perturbed motion (transition process) in relation to the cellular environment of the organism. A set of object of controlled element, which is the cell and the optimal control functions, in parameters such as the binding substance with the cell, the binding rate constant of transfer is the optimum process control system of kinetochore and distribution of drugs into cells. The optimization problem can be interpreted as a two-point boundary value problem, to the solution of which all methods of the calculus of variations are applicable. The law of regulation (control) is presented in its analytical form as a function of the initial coordinates of the control system, i.e. it is necessary to form a differential equation of the function of controlling the movement, penetration of substances directly into the desired cells and distribution in them in accordance with a given optimizing functional. The problem of analytical construction of optimal systems defined in the open domain is studied.

The biological controlled system is considered not in the statistical state, but in their movement and development. Observing systems in motion fundamentally changes the way we study them. In some cases, this allows you to uncover patterns and facts that would otherwise be undisclosed. For example, to assess the performance of many systems or to determine the possibility of their long-term existence, such a functional property of systems as their stability is crucial. To study this property, it is fundamentally necessary to consider the dynamics of processes occurring in systems.

2. Problem statement

We have studied not isolated systems, but systems in the form of cells interacting with each other. It takes into account the diverse connections that are formed between individual parts of cells and determine their properties, behaviour, development, death and reproduction. This approach differs in the relative point of view of the system. This relativity is manifested in the fact that the same set of elements in one case can be considered as an independent system, and in another - as part of some larger system, in which it is included as a component. The properties and features of any objects cannot be properly evaluated and taken into account without considering the diverse relationships and interactions between individual objects and their environment, namely, between cells and blood. This approach is characterized by taking into account the influence of the environment (blood) on the controlled system (cell). In connection with this fact, it is necessary to take into account Nernst's theory of the penetration of matter into cells.

According to the Nernst-Planck model, the membrane is a homogeneous phase with respect to penetrating molecules and ions. However, the ions of many dyes and fluorochromes have amphiphilic properties and, due to this, should be located mainly not in the thickness of the membrane phase, but near its surface (in the interphase). Therefore, it is necessary to take into account the parameters of their binding to the membrane surfaces, i.e. the parameters K and N. And since the two surfaces of the membrane can differ markedly in their structure, in general, the distribution of amphiphilic charged molecules in the membrane can have an asymmetric character.

As a result, only those molecules that are bound on the surfaces are transferred through the membrane. For charged molecules such as fluorescent probes and FS, the flow (F) (in mol/cm^2) of ions through the membrane depends on their concentration in the membrane (r), the mobility of the ions (u), and the gradient of the electrochemical potential (μ) in the membrane:

$$\Phi = -\mathrm{ru}\,\frac{d\mu}{dx}$$

where $\mu = \mu_0 + RTlnC + zF_\Phi\,\varphi$. Here z – molecule's charge, F_Φ – Faraday's number, R – gas constant, φ – potential at a point X.

On the other hand, the diffusion of ions into the cell and their binding to the inner surface of the plasma membrane should be influenced by intracellular membrane systems and, above all, those on which the electric field gradient is formed. In this regard, there is a more general problem – the use of the electrodiffuse model to describe the process of accumulation of charged organic molecules (including FCS) in a living cell, taking into account their interaction with intracellular membranes and organelles.

Effective accumulation of FS in tumor cells is a key factor necessary for the realization of maximum photodynamic damage to the tumor after laser irradiation for the complete recovery of the body. This efficiency manifests itself in damage to the mitochondrial membranes and the cell nucleus, which can be considered as membrane vesicles energetically interconnected through their transmembrane potentials (TMPS).

First, we consider the kinetic process of penetration of PS molecules through the membrane of a single model vesicle.

The penetration of amphiphilic charged PS molecules through the membrane occurs in two phases: incorporation into the membrane at the surface (fast phase) and transfer of molecules through the membrane thickness (slow phase). In this case, some of the molecules bind to the membrane, some of them return to the medium as a result of dissociation, and some of them accumulate inside the membrane vesicle. The kinetic scheme of the penetration of PS molecules of the type of chlorin E6 through the plasma membrane will have the following form:

$$\mathrm{X} + (\mathrm{M_0})_1 \longleftrightarrow (\mathrm{XM_0})_1\, k_{ti} \longleftrightarrow k_{t0}\, (\mathrm{XM_i})_1 \xleftarrow{(K_{di})_1} \xrightarrow[(K_{ai})_1]{} \mathrm{X} + (\mathrm{M_i})$$

where X is a molecule of FS, M_0 and M_i - binding molecules on the surfaces of the membrane, K_a and K_D the rate constants of Association and dissociation of molecules on the surface membrane of this type, and k_0, k_{ti} - a constant of speed of migration of molecules through the membrane inside the vesicles and outside, respectively. The essential point is that only the molecules that occupy the binding centers on the surface of the membrane are transferred.

Then the flow of molecules of FS from outside to inside membrane vesicles (I_0)and from the inside out (I_i) we assume proportional to the concentrations of the bound probe on the outer (r_0) and internal (r_i) surfaces of the membrane and:

$$\Phi_0 = \mathrm{k_0 r_0};\ \Phi_\mathrm{i} = \mathrm{k}_i \mathrm{r_i}.$$

where, k_0 and k_i ⁻ coefficients, depending on membrane's penetration (P) and on gradient of membrane's field (Δφ), then

$$k_i \approx P f(zF_\phi \Delta\varphi).$$

where

$$P = \frac{DK}{d}.$$

where D is the coefficient of diffusion of PS molecules through the membrane, K is the constant of the PS distribution between the membrane and the medium, and d is the thickness of the membrane layer.

In experiments with fluorescent probes, it was shown that for water-soluble amphiphilic charged molecules, the concentration of the bound substance on the membrane surface (r) depends linearly on the concentration of the free substance in the medium (C), the constant (K), and the number of binding centers (N).:

$$r = KNC.$$

Then

$$\Phi_0 = k_{t0} K_0 N_0 C_0;\ \Phi_i = k_{ti} K_i N_i C_i$$

The constants K_0 and K_i actually take into account the transition energy of the substance between the medium - surface phases of the membrane and vice versa.

This energy (ΔG_c) consists of the electric ((ΔG_{el}) and hydrophobic (ΔG_{hd}) energies of the interaction of charged molecules with the membrane surface.

$$\Delta G_c = \Delta G_{el} + \Delta G_{hd}$$

Of course, these properties of substances penetrating into the cell and the properties of the membranes depend on many factors. In particular, it depends on the area in which the tissue or cell is located. I.e., how well oxygen enters this area of the cells. In the tumor zone, there are at least three pools of cells that are significantly heterogeneous in the mitotic index. Then, in the series: normal cells (M_n) - tumor cells in the hypoxia zone (M_{og}) – tumor cells in the zone with normal oxygenation (M_o), the mitotic indices (M) of these cell pools are correlated as follows:

$$M_n << M_{og} < M_o$$

The study of the behavior of a biological system should be carried out taking into account its connections with the environment, i.e. with the surrounding cells and tissues. Since all objects, phenomena and processes are interconnected and affect each other, then, when isolating a cell, it is necessary to take into account the influence of the environment (blood) on this cell and its influence on the environment (blood).

Thus, such a system is characterized by the properties of the objects that make up this system, and the connections of this system with the environment. Its relationship with the environment expressed in terms of specific physico-chemical values that were given (bondage, material flows, capacities, location, etc.).

It is clear that it is impossible to consider all the myriad factors that directly or indirectly affect the behavior of the system. Therefore, we have to take into account the presence of random factors. One of the main features of a controlled system is its ability to change its movement, to move to other states under the influence of control actions.

In the cell system, there is a certain set of possible changes, from which the preferred change is selected. If the system has no choice, then there can be no question of management. Thus, the approach based on the study of the properties and features of the system as a controlled one is characterized.

External influences on the system, which can be disposed of when managing it, are control influences. The effect on the behavior of the cellular system can be achieved both by influencing the rate of penetration of the substance, and by changing its chemical and physical properties, such as lipophilicity or hydrophobicity, as well as changing its parameters of the controlled system-the control object.

The required behavior of the controlled system is achieved through control actions, under the influence of which the system purposefully changes its state. In the case of biological cells to control the rate constants of the mass transfer through the membrane, the binding of the penetrating substance with the cell membrane, the speed of transfer of the substance to achieve the most effective penetration and distribution of the drug in the cell.

Control of an object can be carried out by means of several control actions, each of which is limited by certain limit values. Therefore, in the space of control actions $Y_1, Y_2, \ldots, Y_m$, a region satisfying the conditions can be selected

$$Y_i' < Y_i < Y_i'' \; (i = 1,2,\ldots,m)$$

inside of which there are points representing all possible sets of control actions. This area is called the area of possible impacts.

To control the processes in the cell, you need to change the control actions on it in a certain way. This is realized on the basis of information about the situation that develops during the functioning of the cellular system. Such information can serve as data on the state of the controlled system, its required state, disturbing influences, as well as the characteristics of the controlled system.

Processing this information in the cellular structure according to certain rules allows you to form control actions. Management is necessary not only for the normal functioning of the system, but also to ensure its development in the required direction of recovery of the body as a whole. Management in this case consists in the formation of an effective development of the process of destroying harmful cancer cells and in its implementation. In such a controlled biological system, the problem of stabilization is solved.

3. Stabilization process

The stabilization problem of the system is to maintain the control values of x_i near some constant set values of x_0, provided that there are perturbations that affect the values of x_i.

Based on these concepts, we will further model the tumor with populations of cells that have different mitotic and energetic activities, depending on their location in the tumor and the availability of oxygen, and among them there is a pool of normal cells and necrotic cells.

Suppose there are populations of tumor cells (OK) in the form of different suspensions, simulating the external proliferative zone and the zone of moderate growth (in hypoxia), into which a certain amount of charged PS (chloride) is introduced.

$$\dot{x}_0 = a_{00}x_0 + a_{01}x_1 + a_{02}x_2 + a_{03}x_3$$

$$\dot{x}_1 = a_{10}x_0 + a_{11}x_1 + a_{12}x_2 + a_{13}x_3$$

$$\dot{x}_2 = a_{20}x_0 + a_{21}x_1 + a_{22}x_2 + a_{23}x_3$$

$$\dot{x}_3 = a_{30}x_0 + a_{31}x_1 + a_{32}x_2 + a_{33}x_3$$

where x_i- PS concentration in i^{th} cell, dx_i/dt- substance flows, a_{ij}- constants of the rate of transfer of PS molecules through membranes. Also,

$$a_{ij} = k_{ij}KN_{ij}$$

Next, we will simplify this system and look for solutions for populations of tumor cells in different metabolic states at different stages of the cell cycle, associated with the heterogeneity of these cells in terms of their TMP levels on the outer, mitochondrial and nuclear membranes. For general reasons (for cells that do not have a common medium between the membranes):

$$a_{02} = a_{20} = a_{40} = a_{03} = a_{23} = a_{32} = 0$$

At the same time, we assume the presence of symmetric properties of the outer and inner surfaces of the membranes during the binding of PS molecules. Then the system takes the following form:

$$\dot{x}_0 = a_{00}x_0 + a_{01}x_1$$

$$\dot{x}_1 = a_{10}x_0 + a_{11}x_1 + a_{12}x_2 + a_{13}x_3$$

$$\dot{x}_2 = a_{21}x_1 + a_{22}x_2$$

$$\dot{x}_3 = a_{31}x_1 + a_{33}x_3$$

If we add the control actions, we get:

$$f_\mu(x,t) = y_\mu, \ \dot{y}_\mu = Y_\mu(y,x,t)$$

Here I introduce an algorithm of finding Y_μ:

Let's consider $Y_\mu = b_\mu^\alpha\, y_\alpha$, K_μ^α- const , $\alpha = 0,1,2,3$

$\dot{y}_\mu = b_\mu^\alpha\, y_\alpha$, $y_\alpha(t_0) = f_\alpha(x_0, t_0)$

$$\lim_{t\to\infty} y_\alpha(t) = 0$$

Then a system can be written as follows

$$\dot{y}_\mu = \sum_{\mu,\alpha} b_\mu^\alpha y_\alpha + b_\mu u$$

or

$$\dot{y}_0 = b_0^0 y_0 + b_0^1 y_1 + b_0 u$$

$$\dot{y}_1 = b_1^0 y_0 + b_1^1 y_1 + b_1^2 y_2 + b_1^3 y_3 + b_1 u$$

$$\dot{y}_2 = b_2^1 y_1 + b_2^2 y_2 + b_2 u$$

$$\dot{y}_3 = b_3^1 y_1 + b_3^3 y_3 + b_3 u$$

So for $m = 3,\ f_i(x,t) = y_i,\ i = 0,1,2,3$

$$J = m_{kj} y_k y_j + m\xi^2,$$

Where m_{kj} are the given coefficients of a positive definite quadratic form, m is the given number. The general solution of the equations of the problem in this case will consist of a linear combination of exponential functions of the form:

$$x_k = \sum_{i=0}^{3} \Delta_k(\mu_i) c_i e^{\mu_i t} = \Delta_k(\mu_0) c_0 e^{\mu_o t} + \Delta_k(\mu_1) c_1 e^{\mu_1 t} + \Delta_k(\mu_2) c_2 e^{\mu_2 t} + \Delta_k(\mu_3) c_3 e^{\mu_3 t} + \cdots + \Delta_k(\mu_n) c_n e^{\mu_n t}$$

The desired equation for optimal control of the system of a living cell has the form:

$$\dot{\xi} = \mu_0 \Delta_k(\mu_0) c_0 e^{\mu_o t} + \mu_1 \Delta_k(\mu_1) c_1 e^{\mu_1 t} + \mu_2 \Delta_k(\mu_2) c_2 e^{\mu_2 t} + \mu_3 \Delta_k(\mu_3) c_3 e^{\mu_3 t} + \mu_0 \Delta_{k+1}(\mu_0) c_0 e^{\mu_o t} + \mu_1 \Delta_{k+1}(\mu_1) c_1 e^{\mu_1 t} + \mu_2 \Delta_{k+1}(\mu_2) c_2 e^{\mu_2 t} + \mu_3 \Delta_{k+1}(\mu_3) c_3 e^{\mu_3 t} + \mu_0 \Delta_{k+2}(\mu_0) c_0 e^{\mu_o t} + \mu_1 \Delta_{k+2}(\mu_1) c_1 e^{\mu_1 t} + \mu_2 \Delta_{k+2}(\mu_2) c_2 e^{\mu_2 t} + \mu_3 \Delta_{k+2}(\mu_3) c_3 e^{\mu_3 t} + \mu_0 \Delta_{k+3}(\mu_0) c_0 e^{\mu_o t} + \mu_1 \Delta_{k+3}(\mu_1) c_1 e^{\mu_1 t} + \mu_2 \Delta_{k+3}(\mu_2) c_2 e^{\mu_2 t} + \mu_3 \Delta_{k+3}(\mu_3) c_3 e^{\mu_3 t} + \cdots + \mu_n \Delta_{p+3}(\mu_n) c_n e^{\mu_n t}$$

This "game with the cell" is extremely important for understanding the spread of cancer tumors. Now it is assumed that in cancer cells, a mechanism can be triggered in which the cell, in response to the homeostatic pressure exerted by dividing neighboring cells, abruptly loses its stickiness, detaches from its neighbors and spreads throughout the body-i.e., forms metastases.

But life is life, and the cell, in addition to all this, is also a purely material object: with its mechanical and electrical properties, with its non-equilibrium thermodynamic processes inside it.

Cells require specialized structures to support external mechanical forces. All cells are enclosed in a cell membrane consisting of a thin lipid bilayer that protects the cell from the external environment. Using receptors made up of protein structures, the cell membrane can pass selected molecules into the cell. Inside the cell membrane is the cytoplasm.

4. Conclusion

Biological controlled systems were formed in the process of the evolutionary development of wildlife, for them the purposefulness of management means the influence of the organism on its survival and reproduction.

But the field of application of the theory of management extends to a wide variety of systems: technical, biological, economic, any other, but provided that they are controlled. Accordingly, the application of the general methods of this theory for solving the practical problem of application in the field of biology and medicine is found.

References

[1] . Wei Luo, Rong-Sen Liu,Jian-Guo Zhu, Ying-Chao Li, Hong-Chen Liu. Subcellular location and photodynamic therapeutic effect of chlorin e6 in the human tongue squamous cell cancer Tca 8113 cell line. // Oncology Letters.2014. November 20. P. 551-556.

[2] Askarova K. Z., Morozova G. I., Anoshin A. A. Prediction of PDT efficiency based on the kinetic model of photosensitizer distribution in malignant tumor cells with different levels of transmembrane potentials. // Tez.dokl. 23 International Symposium of the ICO. Dubna, 2016. p. 92.

[3] Astanin S. A., Kolobov A.V., Lobanov A. I., Pimenova T. P., Polezhaev A. A., Solyanik G. I. Influence of spatial heterogeneity of the environment on tumor growth and invasion. Analysis by mathematical modeling methods. Medicine in the mirror of Informatics. Moscow: Nauka, 2008. C.188–223.

[4] Borisov V. A., Zaguskin S. L. / / Oncosurgery, 2009. Vol. 1.. no. 2 .P. 19 -21.

[5] Vereninov A. A., Malakhov I. I. Ion transport in cells in culture. L.: Nauka,1986.

[6] Volobuev A. N. Electrodiffusion of ions through the cell membrane. // Physics of wave processes and radio engineering systems. 2008. T. 11. No. 3. S. 150-156.

[7] Ghanaians S. H., Janukevich M. V., Husnuddin S. M., Ghanaians K. S. Mathematical modeling of the process of tumor growth //Medical Bulletin of Bashkortostan. 2006. T. 1. No. 1. pp. 40-47.

[8] Gelfond M. L. Photodynamic therapy in oncology. //Practical oncology. 2007. Vol. 8. No. 4. pp. 204-210.

IOP Conf. Series: Materials Science and Engineering 1191 (2021) 012016 doi:10.1088/1757-899X/1191/1/012016

Nonlocal theory: modeling of thermomechanical processes in a composite with inclusions in the form of ellipsoids of revolution

Daria Kuvshinnikova

Department of Applied Mathematics, Bauman Moscow State Technical University, 2nd Baumanskaya St., 5, building 1, Moscow, 105005

E-mail: kuvshinnikovadasha@gmail.com

Abstract. An approach to the construction of a mathematical model of a thermomechanical process in a composite material with inclusions in the form of ellipsoids of revolution is considered. This approach is based on models of generalized continuum mechanics and Eringen's nonlocal medium. This approach makes it possible to take into account the processes that occur at the micro- and nanolevel. Physical processes are described using internal state parameters and integral terms, which describe the influence of ellipsoidal inclusions on the propagation of heat and stresses using the "influence function" in the corresponding area. Integro-differential relations are obtained to describe the process of propagation of heat and temperature stresses in a composite material with inclusions in the form of ellipsoids of revolution.

1. Introduction

At present, consolidated structure-sensitive materials are widely used as structural and functional materials for various technical devices, the physical and mechanical properties of which significantly depend on their internal structure. These include, first of all, nanostructured composites with various reinforcement with high-strength fibers: materials modified with particles of various types (carbon nanotubes, fullerenes, etc.) [1 – 3]. The development of a technology for creating nanostructured composites with the ability to predict their thermomechanical characteristics is especially relevant for industries that impose high requirements on the shielding, absorbing, insulating properties of materials used as structural materials with severe restrictions on the mass-dimensional characteristics.

The description of the thermomechanical properties of such materials is possible within the framework of nonlocal models. Such models are based on the methods of generalized mechanics of a continuous medium and got their start in the works of A.K. Eringen [4 – 5]. Further development of the ideas of noncalcous thermodynamics is given in [6 – 9]. The key points in this method are establishing a connection between the characteristics of the micro and macrolevels, as well as taking into account the effects of the spatial nonlocality of the environment.

2. Constitutive relations of nonlocal thermoelasticity

To obtain the governing equations describing thermomechanical processes in a deformable solid, we use the relations of rational thermodynamics of irreversible processes for a medium with internal parameters of a thermodynamic state [6].

The model proposed by A.K. Ehringen [4], is based on the idea that long-range forces that are responsible for the nonlocal behavior of a material at a given point in space $\mathbf{x}$ are adequately described using the distance function $\varphi(|\mathbf{x}' - \mathbf{x}|)$. The function $\varphi(|\mathbf{x}' - \mathbf{x}|)$ is called the "influence function".

Using this assumption, we introduce a relation for the deformations of an anisotropic solid, taking into account the spatial effect of the influence of nonlocality:

$$\varepsilon_{ij}^{(nl)} = p_1\varepsilon_{ij} + p_2 \int\limits_V \varphi(|\mathbf{x}' - \mathbf{x}|)\varepsilon_{ij} d\mathbf{x}' \tag{1}$$

where ε_{ij} are the coefficients of the small strain tensor; $\varphi(|\mathbf{x}' - \mathbf{x}|)$ – influence function that determines the effect of spatial nonlocality, and $\int\limits_V \varphi(|\mathbf{x}' - \mathbf{x}|)d\mathbf{x}' = 1$; p_1, p_2 are the weight fractions of the influence of nonlocality, $p_1 + p_2 = 1$.

The Duhamel-Neumann law for an anisotropic elastic solid has the form [10]:

$$\sigma_{ij} = C_{ijkl}(\varepsilon_{kl}^{(nl)} - \alpha_{kl}^{(T)}\Delta T), \tag{2}$$

where σ_{ij} – stress tensor components; C_{ijkl} – components of the elastic coefficient tensor; $\varepsilon_{kl}^{(nl)}$ – components of the nonlocal deformation tensor; the product $\alpha_{kl}^{(T)}\Delta T$ is responsible for thermal deformations, $\alpha_{kl}^{(T)}$ are the components of the tensor of temperature coefficients of linear expansion; ΔT – absolute temperature increment.

The constitutive relations of the model of thermomechanical processes in a solid can be rewritten in terms of absolute temperature and stresses, if we use the equations of motion and Cauchy relations in the form [10]:

$$\frac{\partial \sigma_{ij}}{\partial x_j} = \rho \frac{\partial^2 u_i}{\partial t^2}, \tag{3}$$

$$\varepsilon_{ij} = \frac{1}{2}\left(\frac{\partial u_i}{\partial x_j} + \frac{\partial u_j}{\partial x_i}\right), \tag{4}$$

where u_i are the components of the displacement vector.

Differentiate equations (1) and (4) twice by t, and the equation (3) by x_j:

$$\frac{\partial^2 \varepsilon_{ij}^{(nl)}}{\partial t^2} = p_1 \frac{\partial^2 \varepsilon_{ij}}{\partial t^2} + p_2 \int\limits_V \varphi(|\mathbf{x}' - \mathbf{x}|)\frac{\partial^2 \varepsilon_{ij}}{\partial t^2} d\mathbf{x}', \tag{5}$$

$$\frac{\partial^2 \varepsilon_{ij}}{\partial t^2} = \frac{1}{2}\left(\frac{\partial}{\partial x_j}\left(\frac{\partial^2 u_i}{\partial t^2}\right) + \frac{\partial}{\partial x_i}\left(\frac{\partial^2 u_j}{\partial t^2}\right)\right), \tag{6}$$

$$\frac{\partial^2 \sigma_{ij}}{\partial x_j^2} = \rho \frac{\partial}{\partial x_j}\left(\frac{\partial^2 u_i}{\partial t^2}\right) \tag{7}$$

Substitute the relations (7) into the equations (6):

$$\frac{\partial^2 \varepsilon_{ij}}{\partial t^2} = \frac{1}{2\rho}\left(\frac{\partial^2 \sigma_{ij}}{\partial x_j^2} + \frac{\partial^2 \sigma_{ij}}{\partial x_i^2}\right). \tag{8}$$

Substitute the obtained relations (8) into the result of differentiation (5):

$$\frac{\partial^2 \varepsilon_{ij}^{(nl)}}{\partial t^2} = \frac{p_1}{2\rho}\left(\frac{\partial^2 \sigma_{ij}}{\partial x_j^2} + \frac{\partial^2 \sigma_{ij}}{\partial x_i^2}\right) + \frac{p_2}{2\rho}\int\limits_V \varphi(|\mathbf{x}' - \mathbf{x}|)\left(\frac{\partial^2 \sigma_{ij}}{\partial x_j'^2} + \frac{\partial^2 \sigma_{ij}}{\partial x_i'^2}\right) d\mathbf{x}'. \tag{9}$$

Differentiate the equation (2) twice with respect to t:

$$\frac{\partial^2 \sigma_{ij}}{\partial t^2} = C_{ijkl}\left(\frac{\partial^2 \varepsilon_{kl}^{(nl)}}{\partial t^2} - \alpha_{kl}^{(T)}\frac{\partial^2 T}{\partial t^2}\right). \tag{10}$$

Substituting the equation connecting stresses and deformations (9) into (10), we obtain the equations of motion in stresses, taking into account nonlocality in space:

$$\frac{\partial^2 \sigma_{ij}}{\partial t^2} = C_{ijkl}\left(\frac{p_1}{2\rho}\left(\frac{\partial^2 \sigma_{ij}}{\partial x_j^2} + \frac{\partial^2 \sigma_{ij}}{\partial x_i^2}\right) + \frac{p_2}{2\rho}\int\limits_V \varphi(|\mathbf{x}' - \mathbf{x}|)\left(\frac{\partial^2 \sigma_{ij}}{\partial x'^2_j} + \frac{\partial^2 \sigma_{ij}}{\partial x'^2_i}\right) d\mathbf{x}' - \alpha_{kl}^{(T)}\frac{\partial^2 T}{\partial t^2}\right). \tag{11}$$

To simplify the relations (11), we multiply the right and left sides by 2ρ, move the time derivatives to the left:

$$2\rho\left(\frac{\partial^2 \sigma_{ij}}{\partial t^2} + C_{ijkl}\alpha_{kl}^{(T)}\frac{\partial^2 T}{\partial t^2}\right) = p_1 C_{ijkl}\left(\frac{\partial^2 \sigma_{ij}}{\partial x_j^2} + \frac{\partial^2 \sigma_{ij}}{\partial x_i^2}\right) + $$
$$+p_2 C_{ijkl}\int\limits_V \varphi(|\mathbf{x}' - \mathbf{x}|)\left(\frac{\partial^2 \sigma_{ij}}{\partial x'^2_j} + \frac{\partial^2 \sigma_{ij}}{\partial x'^2_i}\right) d\mathbf{x}'. \tag{12}$$

Next, on the left side, we will convolve the product of the tensor of elasticity coefficients and the tensor of thermal deformation coefficients by repeating indices, denote $C_{ijkl}\alpha_{kl}^{(T)} = \beta_{ij}$. On the right-hand side of the relation (12), we split the integral into a sum in directions:

$$2\rho\left(\frac{\partial^2 \sigma_{ij}}{\partial t^2} + \beta_{ij}\frac{\partial^2 T}{\partial t^2}\right) = C_{ijkl}\left(p_1\frac{\partial^2 \sigma_{ij}}{\partial x_i^2} + p_2\int\limits_V \varphi(|\mathbf{x}' - \mathbf{x}|)\frac{\partial^2 \sigma_{ij}}{\partial x'^2_i} dV\right) + $$
$$+C_{ijkl}\left(p_1\frac{\partial^2 \sigma_{ij}}{\partial x_j^2} + p_2\int\limits_V \varphi(|\mathbf{x}' - \mathbf{x}|)\frac{\partial^2 \sigma_{ij}}{\partial x'^2_j} d\mathbf{x}'\right) \tag{13}$$

To construct a system of equations describing thermoelasticity in an anisotropic solid, we write down the associated heat conductivity equation in the form [6]:

$$\rho c_\varepsilon \frac{\partial T}{\partial t} = -\frac{\partial q_i^{(nl)}}{\partial x_i} - T C_{ijkl}\alpha_{kl}^{(T)}\frac{\partial \varepsilon_{ij}^{(nl)}}{\partial t}, \tag{14}$$

where ρ is the density of the material; c_ε – specific mass heat capacity at constant deformation; $q_i^{(nl)}$ – components of the heat flow vector.

Next, it is necessary to specify the expression for the heat flux. For an adequate description of the medium model, taking into account spatial nonlocality, we write the relation for the heat flow in the form [7]:

$$q_i^{(nl)} = -\lambda_{ij}^{(T)}\left(p_1\frac{\partial T}{\partial x_j} + p_2\int\limits_V \varphi(|\mathbf{x}' - \mathbf{x}|)\frac{\partial T}{\partial x'_j} d\mathbf{x}'\right). \tag{15}$$

Substituting the relations (1) and (15) in (14), we obtain the heat equation taking into account the nonlocality and connectivity of the temperature and deformation fields:

$$\rho c_\varepsilon \frac{\partial T}{\partial t} = \frac{\partial}{\partial x_i}\left(p_1\lambda_{ij}^{(T)}\frac{\partial T}{\partial x_j} + p_2\lambda_{ij}^{(T)}\int\limits_V \varphi(|\mathbf{x}' - \mathbf{x}|)\frac{\partial T}{\partial x'_j} d\mathbf{x}'\right) -$$

$$-TC_{ijkl}\alpha_{kl}^{(T)}\left(p_1\frac{\partial\varepsilon_{ij}}{\partial t}+p_2\int\limits_V\varphi(|\mathbf{x}'-\mathbf{x}|)\frac{\partial\varepsilon_{ij}}{\partial t}d\mathbf{x}'\right). \tag{16}$$

Equations (13) and (16) are related. The action of mass forces and external loads causes both a displacement field and a temperature field in the body. In turn, the action of heat sources and heating of the body surface are associated with the appearance of a temperature field and a displacement field. Adding the initial and boundary conditions to the equations (13), (16), we obtain a mathematical model describing thermoelasticity in an anisotropic solid.

3. Temperature stresses

In many cases, it is necessary to find a solution for bodies that are only exposed to an external heat source and are heated over the surface. Experiments show that the coupling effect in equations (13) and (16) is insignificant in such cases, therefore, in equation (14), the term $TC_{ijkl}\alpha_{kl}^{(T)}\dot{\varepsilon}_{ij}^{(nl)}$. Then we get a simplified system of equations:

$$2\rho\left(\frac{\partial^2\sigma_{ij}}{\partial t^2}+\beta_{ij}\frac{\partial^2 T}{\partial t^2}\right)=C_{ijkl}\left(p_1\frac{\partial^2\sigma_{ij}}{\partial x_i^2}+p_2\int\limits_V\varphi(|\mathbf{x}'-\mathbf{x}|)\frac{\partial^2\sigma_{ij}}{\partial x'^2_i}d\mathbf{x}'\right)+$$
$$+C_{ijkl}\left(p_1\frac{\partial^2\sigma_{ij}}{\partial x_j^2}+p_2\int\limits_V\varphi(|\mathbf{x}'-\mathbf{x}|)\frac{\partial^2\sigma_{ij}}{\partial x'^2_j}d\mathbf{x}'\right). \tag{17}$$

$$\rho c_\varepsilon\frac{\partial T}{\partial t}=\frac{\partial}{\partial x_i}\left(p_1\lambda_{ij}^{(T)}\frac{\partial T}{\partial x_j}+p_2\lambda_{ij}^{(T)}\int\limits_V\varphi(|\mathbf{x}'-\mathbf{x}|)\frac{\partial T}{\partial x'_j}d\mathbf{x}'\right). \tag{18}$$

Adding the initial and boundary conditions to the equations (17), (18), we obtain a mathematical model describing the temperature stresses in an anisotropic solid, taking into account the effect of nonlocality in space.

4. Nonlocality influence function

In the models proposed by A.K. Ehringen proposed a function of the influence of nonlocality in the form of a Gaussian normal distribution (Figure 1):

$$\varphi(|\mathbf{x}'-\mathbf{x}|)=\frac{1}{(2\pi)^{n/2}}\left(\frac{k}{l}\right)^n\exp\left(-\frac{k^2}{l^2}|\mathbf{x}'-\mathbf{x}|^2\right), \tag{19}$$

where $k=const$ is some constant; l – typical size of the area of influence; n – dimension of the problem. It is important to note that k and l are determined during the experiment.

The choice of the influence function in the form (19) is applicable for mathematical models of consolidated materials, the structural elements of which can be represented in the form of spheres. However, such a representation allows only a narrow class of materials to be described. If the size of inclusions in one direction differs significantly from the other two orthogonal directions, then an ellipsoid of revolution can be taken as an acceptable model describing the shape of such an inclusion.

In the case of elliptical structural elements, the influence function takes place in the form of a normal distribution with characteristic dimensions of the influence area that are different along the orthogonal axes (Figure 2):

$$\varphi(|\mathbf{r}|)=\frac{1}{\sqrt{2\pi}}\left(\frac{k_1k_2}{l_1l_2}\right)\exp\left(-\frac{k_1^2}{l_1^2}|x'-x|^2\right)\exp\left(-\frac{k_2^2}{l_2^2}|y'-y|^2\right), \tag{20}$$

IOP Conf. Series: Materials Science and Engineering 1191 (2021) 012016 doi:10.1088/1757-899X/1191/1/012016

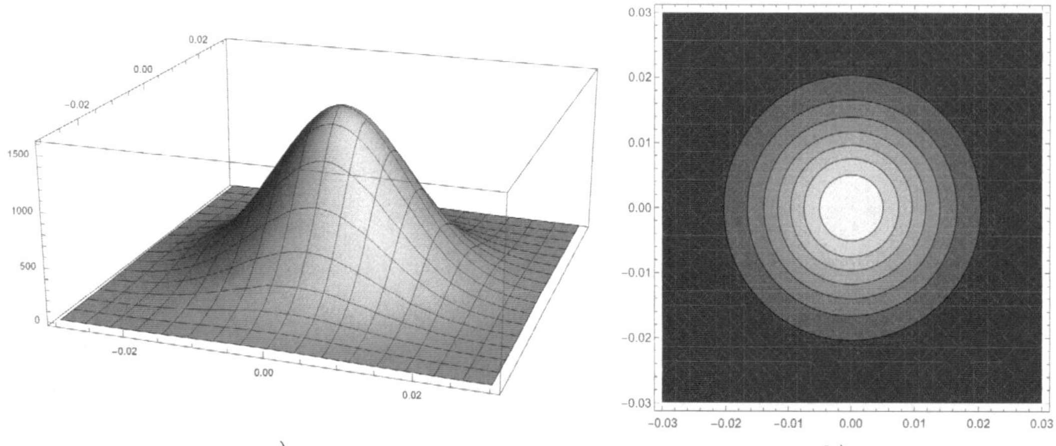

a) b)

Figure 1. Nonlocality influence function for spherical structural elements: a) two-dimensional distribution; b) projection onto a plane.

where k_2, k_2 are dimensional constants; l_1, l_2 – characteristic dimensions of the area of influence in orthogonal directions.

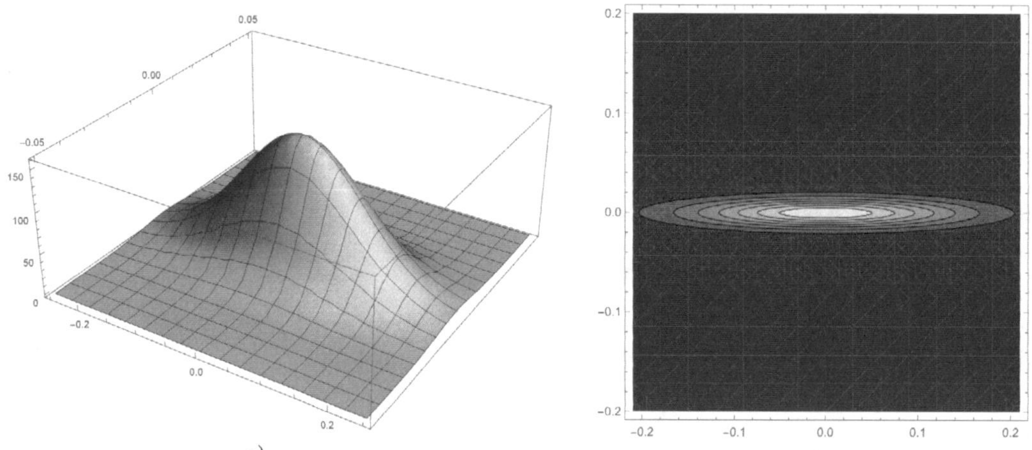

a) b)

Figure 2. Nonlocality influence function for elliptical structural elements: a) two-dimensional distribution; b) projection onto the plane.

Substitution of the influence function in the form (20) into the relations (13), (16) allows you to build a mathematical model of thermoelasticity of a material with structural elements in the form of an ellipsoid of revolution. It should be noted that with an ordered arrangement of inclusions such a material has two mutually orthogonal double axes of symmetry, then the tensor of elasticity coefficients $\hat{\mathbf{C}}$ is characterized by 9 constants due to symmetry, and the tensor of the second rank $\hat{\beta}$ – 3 constants. In this case, it is customary to speak of an orthotropic system [11].

5. Conclusions

A model has been built for describing thermomechanical processes in an anisotropic solid, taking into account the effect of spatial nonlocality. This model is suitable for a variety of structurally sensitive materials, however, in general, the solution of the connected problem of thermoelasticity is rather laborious. In this regard, a simplified model is obtained that describes the temperature stresses in an anisotropic solid and takes into account the spatial nonlocality. The resulting model can be applied to composites with ordered inclusions in the form of spheres or ellipsoids of revolution. The shape of the inclusions is taken into account by introducing the corresponding function of the influence of nonlocality. The paper proposes a function of the influence of nonlocality, based on a multivariate Gaussian distribution, and allowing to take into account inclusions in the form of elongated ellipsoids of revolution. The expressions obtained make it possible to describe thermomechanical processes taking into account the structural features of the material under study.

Acknowledgments
The research was supported by the state assignment of the Ministry of Science and Higher Education of the Russian Federation (project № 0705-2020-0032).

References
[1] Andrievskii R A and Ragulya A V 2005 *Nanostructural Materials* (Moscow, Akademiya)
[2] Alymov M I, Zelensky V A 2007 *Methods of production and physical and mechanical properties of bulk nanocrystalline materials* (Moscow, ELIZ) p 148 (in Russian)
[3] Onami M, Iwashimizu S, Hank K, Shiozawa K and Tanaka K 1987 *Introduction to micromechanics* (Moscow: Metallurgy) p 280 (Moscow, ELIZ) p 148 (in Russian)
[4] Eringen A C 2002 *Nonlocal continuum field teories* (New York-Berlin-Heidelberg: Springer-Verlag) p 393 (Moscow, ELIZ) p 148 (in Russian)
[5] Eringen A C 2006 *Nonlocal continuum mechanics based on distributions* (International Journal of Engineering Science 44(3)) pp 141-147
[6] Zarubin V S, Kuvyrkin G N, Savelyeva I Yu 2013 *Mathematical model of a non-local medium with internal state parameters* (Minsk, Engineering and physical journal, Vol. 86, N 4) pp. 768-773.
[7] Kuvyrkin G N, Savelyeva I Y, Kuvshynnikova D A 2018 *Mathematical model of the heat transfer process taking into account the consequences of nonlocality in structurally sensitive materials* (Journal of Physics: Conference Series, 991(1):012050)
[8] Kuvyrkin G N, Savelyeva I Yu, Kuvshinnikova D A 2019 *Non-stationary heat conduction in a curvilinear plate with account for spatial nonlocality* (Springer New York Consultants Bureau, Journal of engineering physics and thermophysics, Vol. 92. N 3) pp 608-613
[9] Kuvyrkin G N, Savelyeva I Yu and Zhuravsky A V 2018 *Numerical Simulation of Vapor-Phase Epitaxy with Allowance for Diffusion Processes* (Mathematical Models and Computer Simulations, vol 10, issue 3) pp 299–307
[10] Zarubin V S and Kuvyrkin G N 2008 *Mathematical Models of Continuum Mechanics and Electrodynamics* (Moscow: The Bauman University Publishing House) p 512
[11] Novatsky V 1975 *Theory of Elasticity* (Moscow: Mir) p 872

Comparison of some behaviors of planetary mechanism at internal and external gearing of satellite

N I Bondarenko[1], **K B Obnosov**[1] **and A V Panshina**[1]

[1]Bauman Moscow State Technical University, Moscow, 105005, Russia

E-mail: [1]colia.bond@yandex.ru [2]KOB1-Naz@km.ru [3]panalv@mail.ru

Abstract. The potential energy of a planetary type mechanism located in a vertical plane is analyzed for two different configurations with similar parameter values: with internal and external gearing of the satellite. The question of bifurcation of its oscillations is considered.

1. Introduction

Planetary mechanisms can be part of planetary gearboxes, on-board planetary gears of heavy vehicles [1]. In this paper, we study the change in the potential energy [2] of such mechanisms, and hence their movements, when the values of certain characteristic parameters vary.

2. Formulation of the Problem

The considered planetary-type mechanism is located in the vertical plane [3] (Figure 1). Satellite 1 of radius r and mass M_1 can roll without slipping on a fixed cylindrical surface of radius R. Its axis A is attached to the rod 2, whose mass is M_2 and length is L. The rod 2 can rotate around a fixed horizontal axis Oz. In figure 1, the satellite has an internal attachment with a cylindrical surface, and in figure 2- an external engagement. A weightless rod 5 with a load 3 weighing M_3 is rigidly attached to the satellite. The load is fixed on the rod at a certain distance l_3 from the satellite axis. On the carrier 2 at a distance $OD = l_4$ there is a counterweight 4 with a mass of M_4. In the position of the mechanism when $\varphi = 0$ the axis A of the satellite lies on the same vertical and below the axis O of the carrier 2, the longitudinal axis of the rod 5 is located vertically and the load 3 (point B*) is below the axis of the satellite A* (Figure 2).

A mechanism has one degree of freedom. As the generalized coordinate was selected the angle φ (radian) of deflection of the rod 2 from the vertical.

The active forces acting on the mechanism are only the forces of gravity. Let's write out the potential energy of the system:

$$\Pi(\varphi) = g[(M_1 + M_2 + M_3)\cdot(R \mp r) - M_2 L/2 - M_4 l_4]\cdot(1 - \cos\varphi) + \qquad (1)$$
$$+ gM_3 l_3\left[1 - \cos\left(\frac{R \mp r}{r}\varphi\right)\right]$$

The upper " – " sign corresponds to the internal engagement of the satellite (Figure 1), and the lower "+" sign corresponds to the external attachment (Figure 2).

IOP Conf. Series: Materials Science and Engineering 1191 (2021) 012017 doi:10.1088/1757-899X/1191/1/012017

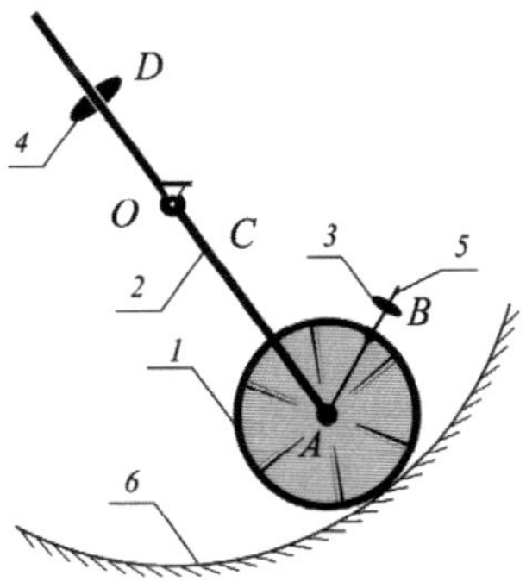

Figure 1. Model of the mechanism with internal attachment.

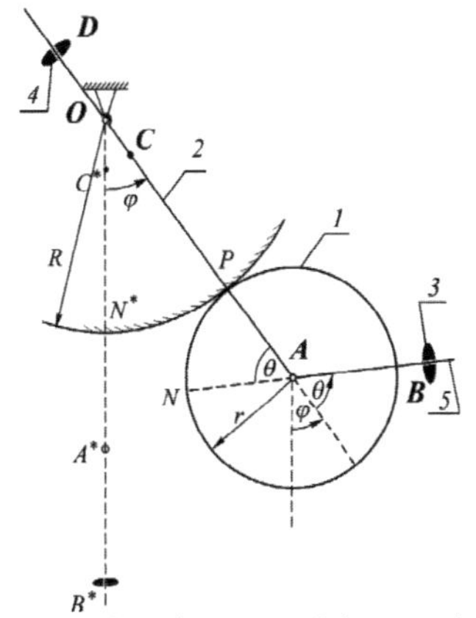

Figure 2. Diagram of the mechanism with external attachment.

3. Investigation of the behavior of the potential energy of the system.

Consider the dependence of the potential energy of the mechanism on the radius of the satellite r for its internal and external engagement. Select the following parameter values:

$$M_1=0.2,\ M_2=0.5,\ M_3=M_4=0.4\ (kg),\ R=0.65,\ l_3=0.2,\ l_4=0.15,\ L=1.8\ (m).$$

Figure 3.a shows graphs of the potential energy at the internal attachment of the satellite, figure 3.b – at the external engagement of the satellite on the segment $\varphi \in [0;8]$ (rad) of the angle of rotation of the rod 2 for some values of the radius r. With both engagements, when the point load 3 is located outside the satellite ($r<<l_3$), $\Pi(\varphi)$ has several extremes on the interval $\varphi \in [0;\pi]$.

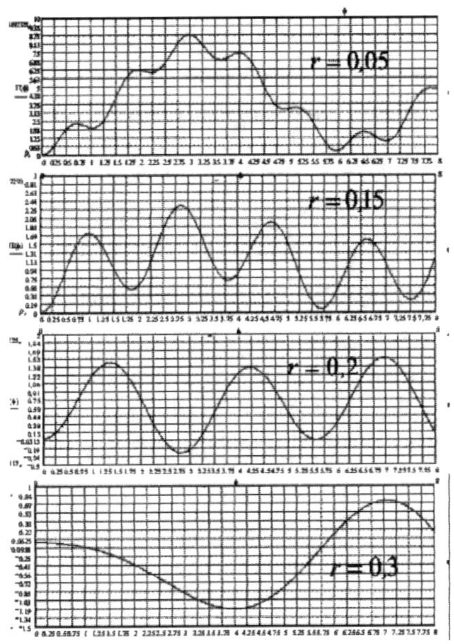

Figure 3.a. Graphs for internal attachment

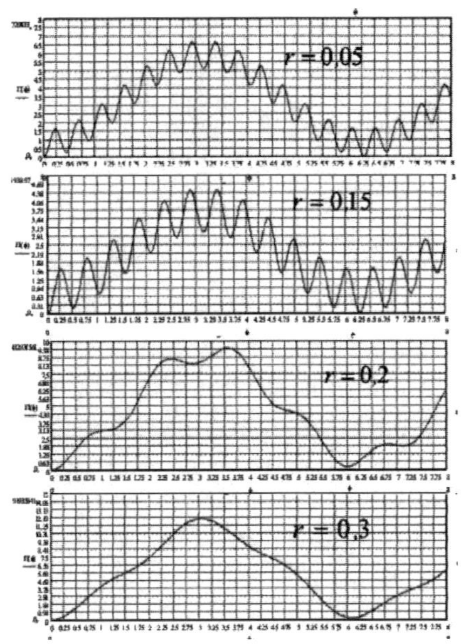

Figure 3.b. Graphs for external attachment

As r increases, the number of extremes decreases. And for the external attachment, their number is greater than for the internal one. This can be explained by the fact that with external engagement, the trajectory of the center of mass C of the satellite is longer. If r=l_3 (load 3 on the rim), the minimum of the potential energy $\Pi(\varphi)$ is only for $\varphi = 0$ and $\varphi \approx 2{,}8$.

If r≈0.29 m for internal engagement the point $\varphi = 0$ changes from the minimum point to the maximum point of the $\Pi(\varphi)$, that is, it becomes an unstable equilibrium position. Near the point $\varphi = \pi$, there is either a maximum or a minimum, which depends on the ratio of radii R and r. Moreover, for internal attachment, in contrast to external attachment, r cannot exceed R.

Figure 4 shows graphs of the potential energy of the mechanism when the satellite is engagement internally. In the first three graphs, the angle of rotation of the rod 2 is taken as follows $\varphi \in [0;128]$ (*rad*). On the other charts $\varphi \in [0;384]$. In figure 5 – external attachment. On the first three graphs, the rotation angle $\varphi \in [0;32]$, on the rest – the angle $\varphi \in [0;128]$ (*rad*).

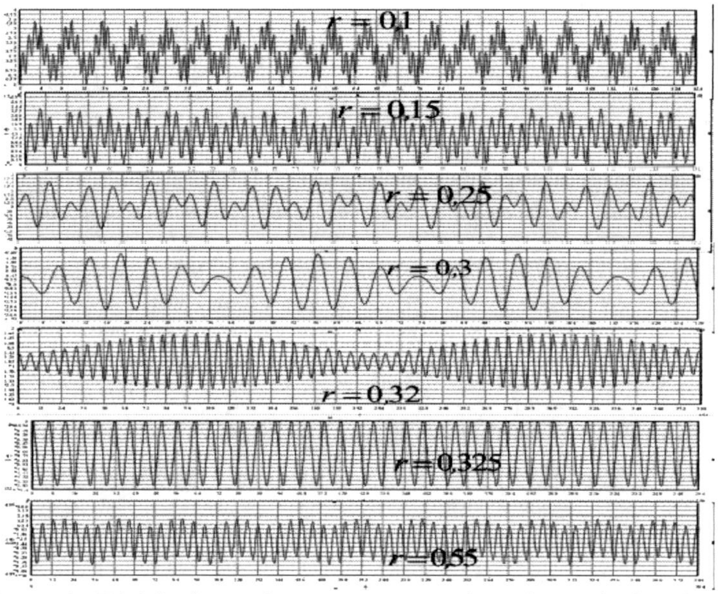

Figure 4. $\Pi(\varphi)$ for internal engagement on a large interval of angle φ.

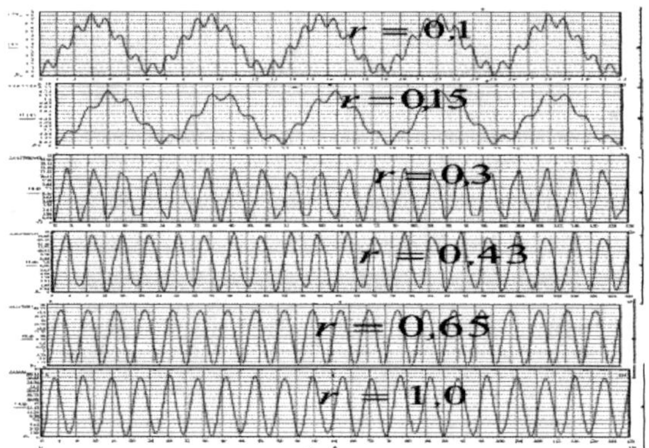

Figure 5. $\Pi(\varphi)$ for external engagement on a large interval of angle φ.

In both cases of satellite attachment, there is a periodicity in the behavior of potential energy. For internal engagement (figure 4), the period increases significantly with increasing r. So for r=0.1 m the period T≈12.5 *rad*, for r=0.32 m the period is T≈204 *rad*. Special situation for $r=R/2$=0.325 m, when $T=\pi$. In this case, the potential energy function (1) takes a simple form:

$$\Pi(\varphi) \approx 1.18\cos\varphi - 0.711.$$

With external attachment (figure 5), the period with increasing r initially increases, but not so significantly: so for r=0.1 *m* the period T≈12.5 *rad*, for r=0.3 the period T≈37.5. And after the period decreases: for r=0.43 T≈12.5, for r=0.65 T≈6.12, but for r=0.1 T≈18.7.

For internal engagement (figure 4), for some values of the radius, for example, at r=0.32, there is a "beating" in the behavior of the function $\Pi(\varphi)$ (1), which is explained by summation cosines with close values of arguments:

$$\Pi(\varphi) \approx 1.44\cos\varphi + 0.78\cos(1{,}031\varphi) - 0.66.$$

In formula (1), for $r \approx \frac{R}{2}$, we have $\frac{R-r}{r} \approx 1$.

With external engagement (figure 5), we do not observe such behavior of the function $\Pi(\varphi)$.

In both cases, a "snake" is observed in the graph of $\Pi(\varphi)$ for certain ratios of radii. This happens when the function reduces to the multiplication of cosines (or cosine and sine) with comparable values of the arguments, but not equal argument values. For external: for r=0.1, r=0.3, r=1. For internal engagement, this is well observed at r=0.1m and r=0.3; r=1.0.

Next, we consider the behavior of $\Pi(\varphi)$ with the internal engagement of the satellite with minor changes in its radius r in the vicinity of the angle $\varphi = 0$. Figure 6 shows graphs of the potential energy in the interval [-1, 1] of the angle φ (*rad*).

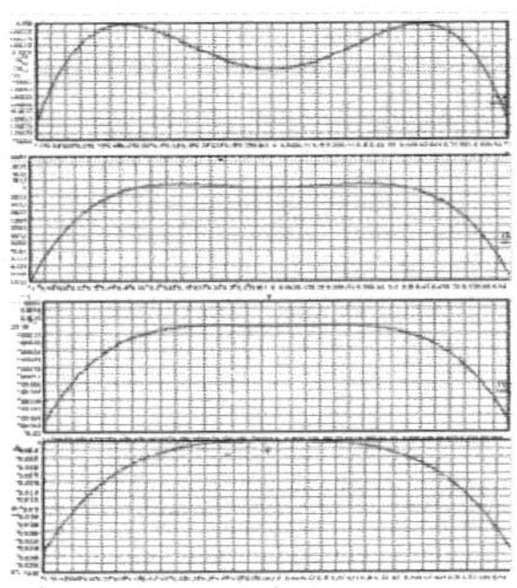

Figure 6. Graph of the function $\Pi(\varphi)$ for internal engagement in the vicinity of a point $\varphi = 0$

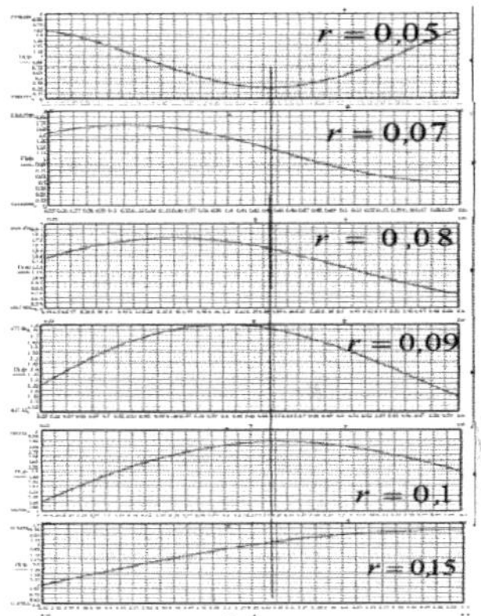

Figure 7. Graph of $\Pi(\varphi)$ for external engagement in the vicinity of a point φ=0.44

Point $\varphi = 0$ is a special point. At r<0.2933, this is the point of stable equilibrium of the mechanism ($\Pi(\varphi)$ has a local minimum). And for r>0.2933, this is the point of an unstable equilibrium position ($\Pi(\varphi)$ has a local maximum). That is, there is a bifurcation – a situation, when, passing through the parameter r^*=0.2933 (*m*) (load 3 inside the satellite), the properties of potential energy change qualitatively. Periodic movements of the system in the vicinity of this point as a result of even minor changes in the size of the satellite become impossible and the system may then behave unpredictably.

Note that the position of stable equilibrium remains for the external engagement at the point $\varphi = 0$ for the considered radius changes (figure 5). A change in the behavior of potential energy $\Pi(\varphi)$ during external engagement can be observed, for example, in the vicinity of $\varphi = 0.445$. For r=0.05 *m*, $\Pi(\varphi)$ has a minimum here (figure 7). After a slight increase in the radius, the second derivative $\Pi''(\varphi)$ becomes negative. So to the right is the position of stable equilibrium of the system. At r≈0.098 m, $\Pi(\varphi)$ has a maximum at this point. At r≈0.098 *m*, $\Pi(\varphi)$ has a maximum at this point.

IOP Conf. Series: Materials Science and Engineering 1191 (2021) 012017 doi:10.1088/1757-899X/1191/1/012017

When r>0.1 m $\Pi''(\varphi) > 0$. This means that to the right of this point is now the position of the unstable equilibrium of the system.

4. Conclusion

Numerical study of the potential energy of the mechanism has shown that even a slight change in the values of its characteristic parameters can significantly change its movement.

References

[1] Kraynev A 1987 Slovar'–spravochnik po mekhanizmam [Dictionary - a directory of mechanisms] (Moscow: Mechanical engineering)
[2] Babakov I 2004 Teoriya kolebaniy [Theory of oscillations] (Moscow: Drofa)
[3] Obnosov K, Bondarenko N and Panshina A 2015 Yestestvennyye i tekhnicheskiye nauki [Natural and technical Sciences] № 10(88) 18-22

IOP Conf. Series: Materials Science and Engineering 1191 (2021) 012018 doi:10.1088/1757-899X/1191/1/012018

Development and research of mathematical models and control algorithms for the separation of bulk materials

Shkapov P.M.[1], Blagoveshchensky I.G.[1]

Blagoveshchensky V.G.[2], Blagoveshchenskaya M.M.[2], Krasnov A.E.[2],

[1]Bauman Moscow State Technical University, Moscow, 105005 Russia
[2]Moscow State University of Food Production 11 Volokolamskoe highway, Moscow, 125080, Russia

E-mail: spm@bmstu.ru; mmb@mgupp.ru;

Abstract. Analysis of the separation process showed that this object is multidimensional, stochastic, nonlinear and has a number of other "inconvenient" properties. All stages of the separation system functioning have a significant impact on the properties of the finished product. Therefore, to improve the separation technology, constant automatic control over all stages of this technological process and an effective control system for this object are required. The creation of such control systems requires the development of methods for their identification, modeling of processes occurring during separation, creation of operational control algorithms implemented by software and hardware systems. At present, the resources for increasing the efficiency of control over the separation of bulk materials by means of automating their main stages have been exhausted, and the way out is seen in the creation and implementation of new generation systems - systems of intelligent functioning and control. The intelligence of the control processes for the separation of bulk materials is determined by the general logic of functioning, as well as the efficiency of making managerial decisions, which is significantly increased if the developed mathematical models and control algorithms that correspond to the separation processes under study are used, which make it possible to evaluate the quality of the decision made on the model. The article provides a literary review of works on the development of methods and models for effective management of production systems based on the use of modern information technologies, which showed that there are many unresolved issues. This work is aimed at solving some of them. The relevance of the topic and the objectives of the research are determined. The research carried out made it possible to develop mathematical models of separation processes of various bulk materials and algorithms for identifying these processes. Based on the results obtained, an algorithm for making decisions by an intelligent software and hardware control system was developed.

1. Introduction

Analysis of the separation process of various bulk materials showed that this object is multidimensional, stochastic, nonlinear and has a number of other "inconvenient" properties. All stages of the separation system functioning have a significant impact on the properties of the finished product. Therefore, to improve the separation technology, an effective control system for this object is required.

The creation of such control systems requires the development of methods for their identification, modeling of processes occurring during separation, creation of operational control algorithms implemented by software and hardware systems. Thus, it becomes possible to move on to the formation of self-adjusting control systems for the technological separation process [1 - 3]. For this, mathematical methods and models are used that are used in the formation of management decisions. With this approach, special attention is paid to the process of mathematical modeling and the conduct of a computational experiment.

The advantage of modeling methods is the ability to get an idea of the functioning of a complex production system under certain conditions by changing the values of the parameters and variables of this system. The modeling is based on the principles of a systematic approach applied to objects [4, 5]. The procedure for choosing the structure and parameters of mathematical models plays an important role [6]. To find models, experimental data are used, on the basis of which, as a result of processing, the model is formed [7-9].

The object identification process goes through several stages. At the first stage, it is necessary to choose the structure of the model based on the collected information about the studied process. At the second stage, it is necessary to choose the criteria for identifying the object and model, which is based on the specifics of the problem. The last stage is to determine the adequacy of the resulting model in accordance with the selected identification criterion. The first identification problem is structural, while the last is parametric [10]. To simulate complex production processes, a "situational approach" [11] is used, which is used in the modeling and management of complex objects [12]. As a rule, the complete situation is formed in accordance with the ideas about the structure of the object under study, about the current state of the control system and about the used control strategies [13]. As a result, any situation is represented by a set of variables that characterize a complex system. However, when studying complex production systems, the exact mathematical solution is not known in advance.

Therefore, approaches based on a combination of mathematical methods with the capabilities of computer modeling are used. In works [14, 15] the foundations of modern methods of system analysis of complex systems are outlined. In the process of development of systems for processing experimental data obtained by automated control systems for technological processes, the analysis of various technological solutions is carried out using models of the objects under study [16,17]. With the help of the obtained models, the influence of various factors on the control process is investigated, effective solutions are formed that improve the functioning of all stages of the separation process of bulk materials [18, 19].

The presence of continuous, continuous-discrete and discrete stages in the separation process makes it necessary to coordinate the stages of the studied production, operating modes, and the composition of technological equipment. The multivariance of the realized technological factors and processing processes necessitates finding optimal technological solutions [20,21].

In papers [22,23], methods of searching for optimal hierarchical structures and mathematical models are considered. Complex production systems form the structure of interconnected objects. The principles of formalizing complex systems are described in [24, 25], where methods for analyzing complex solutions are given, taking into account the dependencies between the used assessment criteria, alternatives and other elements of systems. Systems that contain discrete inputs and are characterized by discrete processing of technological information are classified as discrete systems. Such systems differ in their structural representation, mathematical description and computer implementation [26]. The process of synthesis of complex discrete control systems based on the identification approach, taking into account the characteristics of control quality, is described in the article [27]. For the synthesis of discrete systems, flexible discrete systems are used [28 - 30].

The works of many Russian and foreign scientists are devoted to the development of the theoretical foundations for the management of production systems. A. A. Ashimov, V. N. Volkova, A. A. Emelyanov, J. Clear, D. Mako, M. Mesarovich, G. Simon, I. M. Smirnov, I. Takahara made a significan t contribution to the solution of this problem. , J. Forrester, T. E. Shokhina, V. Z. Yampolsky, etc.

Many scientists have paid enough attention to the development of management methods and models based on the use of modern information technologies. A great contribution to the development of these problems was made by M. Yu. Afanasyev, IV Afonin, M.M. Blagoveshchenskaya, M.I.Bakanov, D.I. Galenko, V.V. Kovalev, A.E. Krasnov, U. U. Oliver, E. A. Utkin, A. V. Tsvetkov and others.

The study of optimization models for the control of production processes based on simulation is considered in the works of P.V. Avdulov, I.G. Blagoveshchensky, A. Yu. Zalozhnev, Yu. Ivashkina, A. A. Kugaenko, I. V. Maksimya, E. A. Nazoykina, A.A. Spirina, V.I. Terekhina, L.L. Terekhova, Ya.M. Urinson, etc.

However, despite the large number of published works, in this area there are still unresolved problems concerning the issues of improving decision-making mechanisms to improve the efficiency of control of separation processes of bulk materials based on the use of modern intelligent technologies, as well as issues of system simulation of the processes under study.

Therefore, an urgent task of increasing the efficiency of such multistage, multidimensional, nonlinear processes such as separation of bulk materials is the modification and adaptation of the methods of mathematical modeling to the processes under study and the development of models and algorithms for controlling this process, which are presented in the form of a system of specialized information support by implemented software and hardware complexes.

At present, the resources for increasing the efficiency of control over the separation of bulk materials by means of automation of their main stages have been exhausted, and we see the way out in the creation and implementation of new generation systems - systems of intelligent functioning and control. The intelligence of control processes for the separation of bulk materials is determined by the general logic of functioning, including a set of intelligently functioning elements and subsystems, as well as the efficiency of management decisions, which is significantly increased if developed mathematical models and control algorithms are used that correspond to the separation processes under study, allowing, with on the one hand, to evaluate the quality of the decision made on the model, and on the other, they are a means of training employees [31].

At the same time, the task of analyzing the studied complex technological processes of separation of bulk materials, the integration of subsystems of individual stages of processing of bulk materials into a united information space, the construction of a control system for the separation process, which makes it possible to form and correct the process under study for each stage of processing, take into account the parameters of actual production is also very urgent.

To create such a system, it is necessary to carry out a mathematical description of the stages of this process. Formalization of complex multistage processes using artificial intelligence technologies requires the development of methods for their identification, the creation of operational control algorithms, implemented by intelligent software and hardware systems. Representation of the characteristics of real technological quantities in the form of a set of inputs, states and outputs allows modeling the main stages of processing the separation process of all types of bulk materials.

The creation of effective intelligent control systems for such objects requires the development of methods for their identification, the creation of operational control algorithms, implemented by intelligent software and hardware complexes.

The purpose of this study is the development and study of mathematical models and algorithms for controlling the separation process of bulk materials, including mechanisms for intellectualizing the main functions of the processes under study.

To achieve this goal in the study, the following main tasks were set and solved: the state and prospects for the development of automated control systems for the separation of bulk materials were analyzed, and actual directions of their intellectualization were formed. Graph models have been developed for the implementation of separation processes of bulk materials using various types of separators, which made it possible to trace the sequence of operations, their structure, patterns, relationships between them and their influence, depending on the equipment used, on the final result.

An experimental study of the separation processes of bulk materials has been carried out, which made it possible to select the parameters that have the greatest impact on the quality of the raw materials obtained after separation, ready for further production of various products. Mathematical models of processes of separation of bulk materials using various types of separators and algorithms for identifying these processes have been developed. An algorithm for controlling the separation of bulk materials, including mechanisms for the intellectualization of the main functions of the processes under study, has been developed.

2. Research

To develop models and control algorithms for the separation of bulk materials, the state and prospects for the development of modern separation processes were analyzed, which, as the analysis showed, is moving towards the intellectualization of all stages of this process and improving the quality of control systems as a whole, expanding their functions and the degree of integration based on modern theoretical and technical base. Separation processes, depending on the stage, proceed in the corresponding units, which are objects of automatic control. To control the separation process, it is necessary to have information about the initial production conditions (characteristics of the bulk product, the state of the unit, etc.), the state of the process itself and its final results [27, 28].

The studies have shown that the solution of the intended problems is possible only if there are mathematical models of this process, which makes it possible to visually study the influence of input controlled and adjustable parameters, disturbing influences on the quality indicators of bulk materials at all stages of preparation of raw materials for production of products, and will also make it possible to predict these technological processes and determine the necessary operating modes of the equipment used.

To develop such mathematical models, experimental research and a survey of experts from various enterprises using bulk materials as raw materials were carried out. Based on the results obtained, structural-parametric and mathematical models of all stages of technological processes of preparing these bulk masses for further production of products were developed, highlighting the factors that have the greatest impact on the main quality indicators.

To solve these problems, the combined effect of technological and mechanical parameters of the separation process on the operating parameters of the mechanical processing of bulk materials was studied. This approach creates conditions for the development of practical solutions for automated control of the investigated processes.

Our studies have shown that separation processes are carried out by various working bodies of technological equipment (sieves, punches, dies, etc.). With obvious differences in the mechanism of processing bulk masses in these processes, the search for a common approach to their study, the choice of the optimal separation scheme is of undoubted interest.

Analysis of separation technological processes on various equipment showed that each of them is an interacting set of operations and transformations intended for the separation of raw materials. These processes are complex in structure, have multidimensionality, some uncertainty, and a random character of parameter changes. In this case, the raw material in the process of separation under the influence of external or internal factors undergoes changes in shape, structure, properties. Therefore, in order to improve the separation process, an important task is to trace the sequence of these transformations in the material under study, their structure, patterns, relationships between them and their influence, depending on the equipment used, on the final result.

3. Graph - models for the implementation of separation processes of bulk materials

In accordance with the task at hand, we will divide each of the considered technological processes into operations. In this case, an operation is understood as a part of the process that is implemented by a functional device or mechanism that provides the necessary interaction of working bodies (or environment) with bulk material to obtain the required result.

Operations are subdivided into main and auxiliary. The main ones are those that result in the processing of the material, which changes its properties in accordance with the quality requirements.

Auxiliary operations are those, the implementation of which does not change the properties of the material. These are, for example, operations of loading, transportation, orientation of the investigated materials in space and in time, the withdrawal of material from the working bodies, etc.

However, in a comprehensive study of separation processes, there is no concept of secondary operations, since the exclusion or insufficient attention to any link violates the integrity of the process.

The number of main and auxiliary operations determines the minimum set of working bodies and functional devices for the implementation of the studied separation process.

With a detailed differentiation of technological processes in order to identify their structure, it is very important to know the characteristics of the connections between their elements. The intertwining connections between elements are such that the slightest change in any one of them or its parameter entails a change in the characteristics of many other connections, and also leads to the exclusion of some of them or the emergence of new ones.

To solve this problem, it is advisable to use the method of mathematical modeling using the theory of finite directed graphs [27, 28]. This method allows you to apply formal methods of analyzing systems, to build visual models of various technological separation processes carried out on the corresponding separators, which is convenient for displaying the complex structure of the system, logical and temporary connections between the elements of the system.

Figure 1 shows a graph - a model of the implementation of the process of separating materials using rectilinear vibrations of sieves, for example, on separators of the ZSM-10, SM-50 and ZSM-100 types.

When constructing these models, the source of the X_0 graph was identified with the input source material, the final vertex of the X_9 and X_8 graphs characterizes the end of the operation and the receipt of material with the specified parameters, the arc (for example, U_{21}, U_{98}, etc.) - the execution of operations; its own contour (loop) (for example, X_{33}) expresses the idle cycles of a device or mechanism, as a result of which the properties of the material will not be changed, i.e. expresses the loss when performing an operation, arcs such as U_{10}, U_{20}, etc. characterize operations for the withdrawal of waste material.

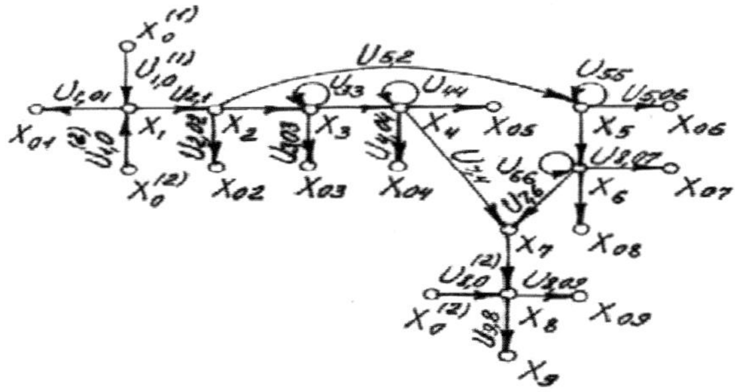

Figure 1. Graph-model of the implementation of the separation process of bulk materials using rectilinear vibrations of sieves

Since in the resulting graph model, shown in Figure 1, a connection is established between the set of independent variables $X_0 \equiv \{x_0^{(1)}, x_0^{(2)}, ...\}$ and many dependent variables $X_i \equiv \{X_1, X_2, X_3, ..., X_L\}$, then the graph G = (X,U) (Figure 1) can be described by a system of equations of the following form:

$$
\begin{aligned}
&X_1 = a_{1,0}^{(1)} X_0^{(0)} + a_{1,0}^{(2)} X_0^{(2)} - a_{1,01} X_{01};\\
&X_2 = a_{2,1} X_1 - a_{2,02} X_{02};\\
&X_3 = a_{4,3} X_2 + a_{3,3} X_3 - a_{3,03} X_{03};\\
&X_4 = a_{1,0} X_3 + a_{4,4} X_4 - a_{4,04} X_{04} - a_{4,05} X_5;\\
&X_5 = a_{5,2} X_2 + a_{5,5} X_5 - a_{5,06} X_{06};\\
&X_6 = a_{6,5} X_5 + a_{6,6} X_6 - a_{6,07} X_{07} - a_{6,08} X_{08};\\
&X_7 = a_{7,4} X_4 + a_{7,6} X_6;\\
&X_8 = a_{8,7} X_7 + a_{8,0}^{(2)} X_0^{(2)} - a_{8,09} X_{09};\\
&X_9 = a_{9,8} X_8
\end{aligned}
\tag{1}
$$

The system of equations (1) displays a diagram of the implementation of the separation process of bulk materials using rectilinear vibrations of sieves and using separators of the ZSM-Yu, ZSM-50, ZSM-100 types, where $a_{1,0}^{(1)}, a_{2,1}, ..., a_{LL}$ - the coefficients of the system of equations reflecting the operations of the separation process (implemented by the working bodies of the equipment) are expressed in the graph model by the corresponding arcs $U_{1,0}^{(1)}, U_{2,1}, ..., U_{LL}$ between the variables Xi and Xj, both dependent and independent.

In the system of equations (1) $x_j = a_{ji} x_i \equiv U_{ji} x_i$, where $U_{ji} \equiv a_{ji}$ - an arc is identified with a numerical value or a coupling function and is called a transfer ratio or transfer function.

The analysis of the graph model presented in Figure 1 made it possible to identify the main properties of the adjacency matrix and calculate the coefficients of asymmetry K_a, flow rates $K_{пот}$ and indicators of the presence of own contours in the main flow direction of the graph $K_{ск}$. The numerical values of these coefficients make it possible to judge the rationality of the structure of the process under study, its flow rate and progressiveness of the scheme under consideration [29]. The graph asymmetry coefficient characterizes the flow rate of the used process scheme and is determined from the expression:

$$K_a = \frac{\sum \overleftarrow{m}_i^j}{\sum \vec{m}_i^j}, \tag{2}$$

Where $\sum \overleftarrow{m}_i^j$ - the members of the adjacency matrix located below the diagonal are numerically equal to the number of arcs outgoing from the vertices X_j, and entering the vertex X_i;

$\sum \vec{m}_i^j$ -terms of the adjacency matrix located on top of the diagonal reflect the arcs of the graph in the forward direction, i.e. the main flow of implementation.

For the graph model shown in Figure 1

$$K_a = 0{,}453 \tag{3}$$

The smaller the K_a value, the less the graph asymmetry, other things being equal. When analyzing the processes of preparation of bulk materials, K_a characterizes the rationality of the separation process carried out with the help of certain specified equipment.

The value of the coefficient of the flow rate of the process $K_{пот}$ and the indicator of the presence of its own contours in the main flow of the direction of the graph $K_{ск}$ are also the most important indicators of the progressiveness and rationality of the technological process [30].

These indicators characterize the possibility of using automation tools and intelligent technologies to control the studied separation processes. $K_{пот}$ and $K_{ск}$ were determined for the graph model shown in Figure 1

$$K_{пот.} = 0{,}5462, \qquad K_{ск} = 0{,}5823. \tag{4}$$

In addition to separators with rectilinear vibrations of sieve housings, separators with circular translational vibrations of sieve housings, as well as air separators, are widely used [31]. Figure 2 shows a graph-model of the implementation of the separation process of bulk materials using circular translational vibrations of sieve housings.

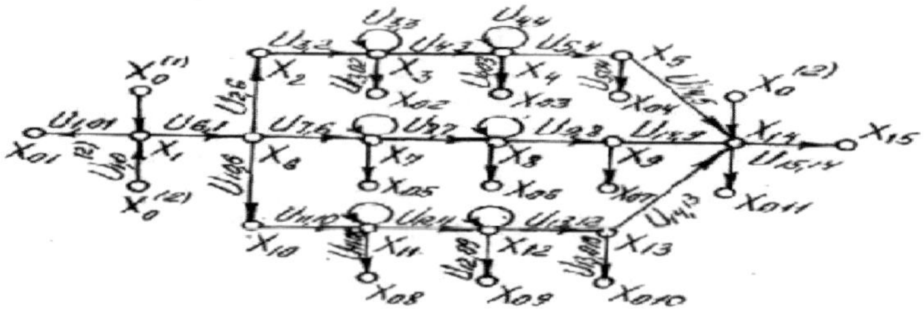

Figure 2. Graph-model of the implementation of the separation process of bulk materials using circular translational vibrations.

The process of separation of bulk materials using circular translational vibrations it is carried out, for example, on separators of the ZSSh, ZRSh, ZRM or ZRL type.

Obtained in Figure 2, the graph of the implementation of the separation process can be described by a system of equations of the form:

$$\begin{aligned}
&X_1 = a_{1,0}^{(1)} X_0^{(0)} + a_{1,0}^{(2)} X_0^{(2)} - a_{1,01} X_{01};\\
&X_2 = a_{2,6} X_6;\\
&X_3 = a_{3,2} X_2 + a_{3,3} X_3 - a_{3,02} X_{02};\\
&X_4 = a_{4,3} X_3 + a_{4,4} X_4 - a_{4,03} X_{03};\\
&X_5 = a_{5,4} X_4 - a_{5,04} X_{04};\\
&X_6 = a_{6,1} X_1;\\
&X_7 = a_{7,6} X_6 + a_{7,7} X_7;\\
&X_8 = a_{8,7} X_7 + a_{8,8} X_8;\\
&X_9 = a_{9,8} X_8\\
&X_{10} = a_{10,6} X_6\\
&X_{11} = a_{11,10} X_{10} + a_{11,11} X_{11} - a_{11,08} X_{08};\\
&X_{12} = a_{12,11} X_{11} + a_{12,12} X_{13} - a_{12,09} X_{09};\\
&X_{13} = a_{13,12} X_{12} - a_{13,010} X_{010};\\
&X_{14} = a_{14,9} X_9 + a_{14,5} X_5 - a_{14,13} X_{13} + a_{14,0} X^{(2)} - a_{14,011} X_{011}\\
&X_{15} = a_{15,14} X_{14}
\end{aligned} \tag{5}$$

Based on the resulting graph model (Figure 2), an adjacency matrix was built and the following were determined:

$$K_a = 0{,}2581;\ K_{пот.} = 0{,}7419;\ K_{ск.} = 0{,}3216. \tag{6}$$

Figure 3. shows a graph-model of the implementation of the separation process of bulk materials using air separation. The process is carried out, for example, on separators such as BDA, BAS.

Based on the resulting graph model (Figure 3.), an adjacency matrix was built and calculated:

$$K_a = 0{,}3764;\ K_{пот.} = 0{,}6236;\ K_{ск.} = 0{,}4984. \tag{7}$$

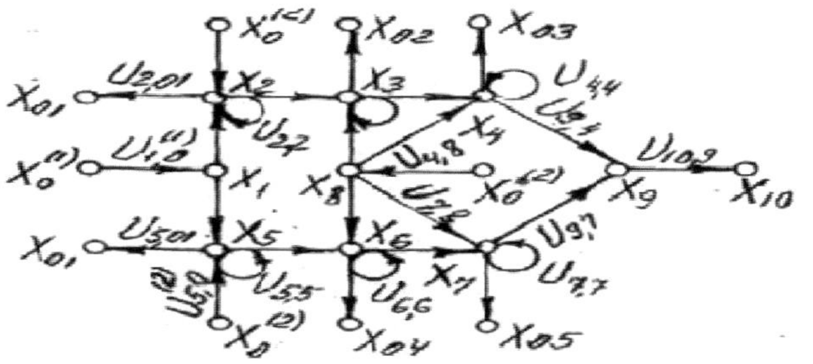

Figure 3. Graph-model of the implementation of the separation process of bulk materials using air separation

The graph shown in Figure 3 can be described by a system of equations of the form:

$$\begin{aligned}
X_1 &= a_{1,0}^{(1)} X_0^{(1)};\\
X_2 &= a_{2,0} X_0^{(2)} + a_{2,2} X_2 + 2_{3,1} X_1 - a_{2,01} X_{01};\\
X_3 &= a_{3,2} X_2 + a_{3,3} X_3 - a_{3,02} X_{02} + a_{8,3} X_8;\\
X_4 &= a_{4,3} X_3 + a_{4,4} X_4 + a_{4,8} X_8 - a_{4,03} X_{03};\\
X_5 &= a_{5,1} X_1 - a_{5,0}^{(2)} X_0^{(2)} + a_{5,5} X_5 - a_{5,01} X_{01}\\
X_6 &= a_{6,5} X_5 + a_{5,0} X_8 + a_{6,6} X_6 - a_{6,04} X_{04};\\
X_7 &= a_{7,6} X_5 + a_{8,6} X_8 + a_{6,6} X_6 - a_{7,05} X_{05};\\
X_8 &= a_{8,0}^{(2)} X_0^{(2)};\\
X_9 &= a_{9,4} X_4 + a_{9,7} X_7\\
X_{10} &= a_{10,9} X_9
\end{aligned} \tag{8}$$

The graph models obtained above for the implementation of separation process flow diagrams in various ways (Figures 1 - 3) and the corresponding systems of equations (1, 5 and 8) express the models of changes in the properties of the materials under study as a result of a complex of operations performed by separator devices and mechanisms.

Figure 4. shows a graph-model of the implementation of the separation process of bulk materials, carried out in air sieve cleaning and sorting machines.

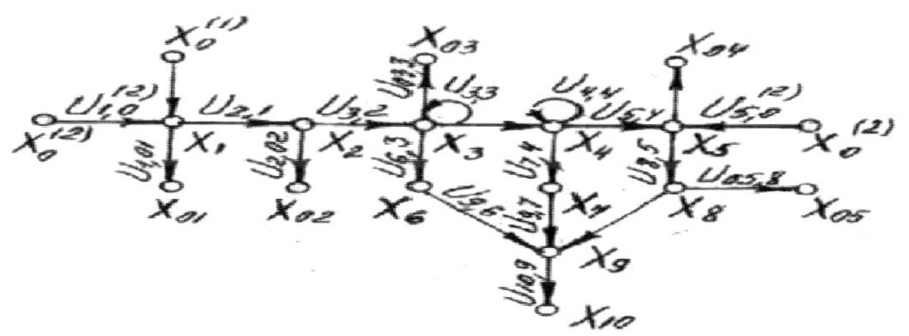

Figure 4. Graph-model of the implementation of the process of the initial bulk material using air sieve cleaning and sorting machines

The resulting graph model was used to build an adjacency matrix and determine:

$$K_a = 0{,}1673;\ K_{пот.} = 0{,}8327;\ K_{ск.} = 0{,}3649. \tag{9}$$

The resulting graph model can be described by a system of equations of the following form:

$$\begin{aligned}
X_1 &= U_{1,0}^{(1)} X_0^{(1)} + U_{1,0}^{(1)} X_0^{(1)} - U_{01,1} X_{01};\\
X_2 &= U_{2,1} X_1 - U_{02,2} X_{02};\\
X_3 &= U_{3,2} X_2 + U_{3,3} X_3 - U_{03,3} X_{03};\\
X_4 &= U_{4,3} X_3 + U_{4,4} X_4;\\
X_5 &= U_{5,4} X_4 + U_{5,00}^{(2)} X_0^{(1)} - U_{04,,5} X_{04};\\
X_6 &= U_{6,3} X_3;\\
X_7 &= U_{7,4} X_4;\\
X_8 &= U_{8,5} X_5 - U_{05,8} X_{05};\\
X_9 &= U_{9,6} X_6 + U_{9,7} X_7 + U_{9,8} X_8;\\
X_{10} &= U_{10,9} X_9
\end{aligned} \tag{10}$$

The resulting system of equations (10) displays a diagram of the implementation of the separation process of the initial bulk material using air sieve cleaning and sorting machines.

The conducted research and the results obtained made it possible to apply formal methods of analyzing systems, to build visual models of the separation technological processes discussed above, carried out on various types of separators, to assess the rationality of the schemes and machine structures used for separation, and the possibility of obtaining an idea of the functioning of a complex production system under certain conditions for by changing the values of parameters and variables of this system.

The modeling is based on the principles of a systematic approach applied to separation objects [32]. An important role is played by the procedure for choosing the structure and parameters of mathematical models [33]. To find models, experimental data are used, on the basis of which, as a result of processing, a model is formed [34].

4. Structural and parametric analysis of technological processes for the separation of bulk materials

To conduct a structural and parametric analysis of the separation processes of bulk materials and the development of appropriate mathematical models, experimental studies of the studied technological system were carried out and a structural-parametric model of relationships between the parameters of state and quality at the stage of the separation of bulk materials with the release of factors that have the greatest influence on the efficiency of cleaning was developed the bulk material coming out of the separator.

As a result of a survey of experienced specialists - experts at enterprises, the most informative parameters of the separation process of bulk masses were selected and matrices of relationships between them were compiled (Table 1).

Table 1

The most informative parameters of the separation process

Stage of the process of preparing bulk material for separation	**Name controlled technological parameter**	**Range changes**	**Unit Measurements**
Separation of bulk material	$N_к$ (X_1) – sieve vibration frequency	775-785	1 minute
	$Q_з$ (X_2) - consumption of bulk material	21-27	ton / hour
	$Q_в$ (X_3) – air flow	7900-8100	m^3 / hour
	Mc (X_4) – material moisture	9-14	%
	Mв (X_5) – air humidity	30-90	%
	Tв (X_6) – air temperature	16-35	^0C
	Do (X_7) – hole diameter of bottom sieves	6,5-8	millimeters
	Nв (X_8) – sieve rotation frequency	105-109	1/ minute
	Rк (X_9) – sieve vibration amplitude	5-7	millimeters
	Yk (Y_1= X_{10})– cleaning efficiency of bulk material	75-85	%

Table 2, in the form of a matrix of expert assessments, presents the results of a survey of experts on the informativeness of the parameters that have the greatest impact on the quality indicators of bulk materials during the separation process at the stage of preparation for further production. The symbol (•) denotes the presence, in the opinion of experts, of a functional connection between the input and output variables and the state parameters of a particular technological process.

Table 2.

Matrix of expert assessments of relationships between parameters at the stage of separation of bulk material

	X_1	X_2	X_3	X_4	X_5	X_6	X_7	X_8	X_9	X_{10}	X_{11}	X_{12}
X_1	1											

X_2	•	1	•				•	•	•	•	•	•
X_3			1	•			•	•	•	•		•
X_4		•	•	1	•	•	•	•	•	•	•	•
X_5			•		1					•		•
X_6			•	•		1				•		•
X_7	•	•	•				1	•	•	•	•	•
X_8		•	•					1		•	•	•
X_9		•	•				•	•	1	•	•	•
X_{10}	•	•	•	•	•	•	•	•	•	1		
X_{11}	•	•	•	•			•	•	•	•	1	•
X_{12}	•	•	•	•	•	•	•	•	•	•	•	1

5. Results and discussion

For a more accurate mathematical analysis of the quality indicators of bulk masses, it is necessary to construct structural - parametric and mathematical models of the separation process. The development of such models is associated with their representation in the form of a system of interacting elements and as a subsystem of some external system [35].

Our experimental studies and the statistical data obtained at the enterprises allowed us to identify the main input parameters and disturbing factors that affect the course of the process of separating bulk materials. The obtained experimental results were compared with the expert testimony, on the basis of which a structural - parametric matrix of the functioning of the connections between the studied parameters at the stage of separation of bulk materials was constructed (Table 2).

The development of structural and parametric models was carried out in several stages. At the first stage, the presence of links between the parameters was established using the correlation coefficients according to formula (11) and a table of correlation links was formed that reflects the depth of the statistical relationship between the parameters at all stages of separation.

$$r_{ij} = \frac{\frac{1}{n-1}\sum_{k=1}^{n}(\bar{x}_i - x_{ki})(\bar{x}_j - x_{kj})}{\sqrt{S_{x_i}^2}\sqrt{S_j^2}} \quad , \qquad (11)$$

где: r_{ij} –correlation relationship between the i and j parameters;;

$\bar{x}_i, \bar{x}_j$ – the average value of the i and j parameter;

$S_{x_i}^2, S_{x_j}^2$- variance of the i and j parameters.

The significance of the coefficients r_{ij} was checked according to the formula (12) (Student's test) so that at $t \geq t_{\text{кр}}$ the coefficient was taken to be significant,

$$t_{ij} = \frac{r_{ij}\sqrt{n-2}}{\sqrt{1-r_{ij}^2}} > t_{\text{кр}}^{v,\alpha} \qquad (12)$$

otherwise, the coefficient r_{ij} was equal to 0.

At the second stage, the nature of the relationship between the parameters was determined by the method of regression analysis. Separating the groups of strongly related parameters ($r_{ij} \geq$ 0,5) line by line using formula (13), the coefficients of linear multiple regression were calculated for each parameter:

$$x_i = p_{i0} + \sum_{\substack{j=1 \\ j \neq i}}^{m} p_{ij} x_j; \qquad i = \overline{1,n}, \qquad (13)$$

где $p_{i0}, p_{i1}, p_{i2} \dots, p_{im_j}$ – коэффициенты связи j-ого параметра ($j = \overline{1, m_i}$) с i-м.

Regression coefficients were calculated using the least squares method based on the log of observations of controlled parameters in n experiments ($k = \overline{1,n}$).

$$R(\vec{p}) = \sum_{k=1}^{n}(x_k - \bar{x}_k)^2 \to min \qquad (14)$$

The essence of the method is that the sum of the squares of the deviations of the regression model from the studied indicator is min. Passing to the increments Δx, for each row of the correlation matrix in general form, we have:

$$\Delta x_i = \sum_{k=1}^{n} p_{ij} \Delta x_j; \; i = \overline{1,m} \qquad (15)$$

where m_i – is the number of factors Δx_j, that have a sufficiently strong correlation with the i deviation Δx_i.

Procedure for finding the coefficients p_j of multiple regression $y = p_0 + \sum_{j=1}^{n} p_j x_j$ with a single repetition of the experiments, it includes the calculation of the coefficients of the system of equations (formula 16) in the form of arrays a_{ij} (formula 17) and b_i (formula 18).

$$\sum_{k=1}^{n}\sum_{j=0}^{m_i} p_j \cdot x_{kj} x_{ki} = \sum_{k=1}^{n} y_k \cdot x_{ki};\ i = \overline{0, m_i} \tag{16}$$

$$a_{ij} = \sum_{k=1}^{n} x_{ki} \cdot x_{kj};\ ij = \overline{0, m} \tag{17}$$

$$b_i = \sum_{k=1}^{n} x_{ki} y_k;\quad i = \overline{0, m} \tag{18}$$

Next, the system of equations $\sum_{j=0}^{m} p_j \cdot a_{ij} = b_i; i = \overline{0, m}$ is solved with respect to the unknown coefficients p_j; $(j = \overline{0, m})$ by the Gauss – Jordan method [36].

The obtained coefficients $p_j = a_{j,m+1}; j = \overline{0, m}$ are tested for adequacy according to the Fisher criterion (Formula 19).

$$F = {S_y^2}/{S_{ad}^2}\ , \tag{19}$$

where
$$S_y^2 = \frac{1}{n-1}\sum_{k=1}^{n}(\bar{Y} - y_k)^2;$$
$$S_{ad}^2 = \frac{1}{n-1}\sum_{k=1}^{n}\left(\sum_{j=1}^{m} p_j x_{kj} - y_k\right)^2.$$

In order for the coefficients of the influence of parameters of different physical nature to be compared with each other, it is necessary to move from the matrix of dimensional coefficients P_{ij} to the matrix of dimensionless, relative characteristics of connections.

For this, the linear relationship formula (15) considers the relative values $\Delta x_i^* = \frac{\Delta x_i}{\Delta x_i^0}$ in fractions of the permissible deviation Δx_i^0 or relative to the standard deviation σ_{x_i} (formula 20).

$$\Delta x_i^* = \frac{x_i - \bar{x}_i}{\sigma_{x_i}} = \frac{\Delta x_i}{\sigma_{x_i}} \tag{20}$$

Thus, the original regression equation (Formula 15) takes the form:

$$\sigma_{x_i} \Delta x_i^* = \sum_{\substack{j=1 \\ j \neq i}}^{n} p_{ij} \cdot \frac{\sigma_{x_j}}{\sigma_{x_i}} \tag{21}$$

with dimensionless coefficients (formula 22), according to which, in the future, a mathematical model of the dependence of quality indicators on parameters is built.

$$C_{ij} = p_{ij} \cdot \frac{\sigma_{x_j}}{\sigma_{x_i}};\ i = \overline{1, n} \tag{22}$$

Taking into account the above studies and the identified most informative parameters affecting the quality of bulk materials in the separation process, a structural-parametric model of the separation process of bulk materials using a separator with circular translational vibrations of sieves was developed, since for this method the most optimal values of the main asymmetry indicators K_a, flow rate $K_{пот}$ and the presence of their own contours in the main stream $K_{ск}$ (Figure 5).

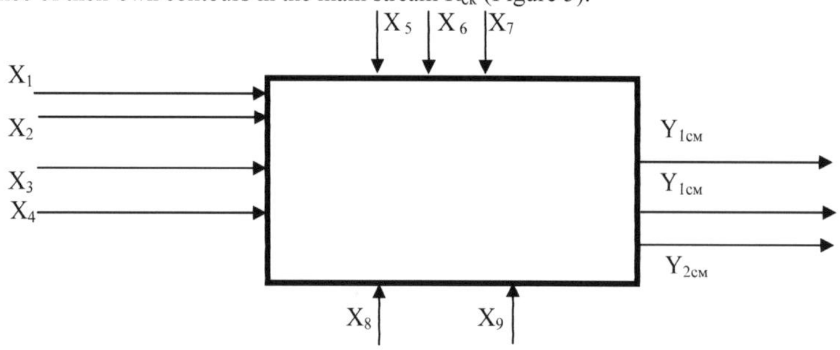

Figure 5. Structural - parametric model of the separation process of bulk materials using a separator with circular translational vibrations of sieves

Input parameters:
N_k (X_1) – sieve vibration frequency, 1 / min– 775-785
$Q_з$ (X_2) – consumption of separated material, tons / hour – 21-27

$Q_в$ (X_3) – расход воздуха, м^3/час – 7900-8100
M_c (X_4) – moisture content of separated masses, % - 9-14
$M_в$ (X_5) – air humidity, % - 30-90
$T_в$ (X_6) – air temperature, ^{0}C -16-35
D_0 (X_7) – hole diameter of bottom sieves, mm - 6,5-8
$N_в$ (X_8) – sieve rotation frequency, 1 / min – 105-109
$R_к$ (X_9) – sieve vibration amplitude, mm – 5-7
Output parameters: $Y_{1сп}$ (X_{10}) - coefficient of impurity extraction (cleaning efficiency), % - 75-85
$Y_{2сп}$ (X_{11})– particle size of separated masses, mm
$Y_{3сп}$ (X_{12})– mass moisture after cleaning, %.

Structural and parametric modeling of the separation process was reduced to the construction of matrixes of relationships between the grouped parameters. In this case, the main task was to find comparable characteristics of the relationships between the parameters of the state of the technological system with the subsequent construction of a situational model of the state of the process and the algorithmization of procedures for its identification and forecasting.

Based on the obtained characteristics of the relationships between the parameters, a quality criterion was selected that meets the current requirements for controlling the separation process. In general terms, the criterion is written as follows:

$$Y_k = C_1X_1 + C_2X_2 + C_3X_3 + \ldots + C_nX_n, \quad (23)$$

On the basis of the calculations, mathematical models of the separation process of bulk materials using a separator with circular translational vibrations of sieves were obtained:

$$Y_1 cn = 0,65\, X_1 + 0,36\, X_2 + 0,56\, X_3 + 0,68\, X_4 + 0,41\, X_7 + 0,55\, X_8 + 0,54\, X_9 \quad (24)$$

$$Y_2 cn = 0,25\, X_1 + 0,76\, X_7 + 0,68\, X_8 + 0,57\, X_9 \quad \ldots\ldots\ldots\ldots \quad (25)$$

$$Y_3 cn = 0,15\, X_5 + 0,11\, X_6 \quad \ldots\ldots\ldots \quad (26)$$

Thus, the proposed method of structural-parametric modeling makes it possible to determine the influence of the input parameters on the output indicators of the quality of bulk materials after separation.

Based on the obtained characteristics of the relationships between the parameters, a quality criterion $Y_к$, was selected that meets the current requirements for the separation process of bulk materials on a separator with circular translational movements, which made it possible to determine the most informative parameters and find the optimal quality criterion for the process under study:

$$Y_К = 0,86\, Y_{1сп} + 0,76Y_{2сп} - 0,18Y_{3сп} + 0,12Y_{4сп} + 0,05Y_{5сп} \quad (27)$$

6. Conclusions

An experimental study of the separation process of bulk materials was carried out, which made it possible to select the parameters that have the greatest impact on the quality of the finished product obtained after separation.

Graph-models have been developed for the implementation of separation processes of bulk materials, which made it possible to trace the sequence of operations, their structure, patterns, relationships between them and their influence, depending on the equipment used, on the final result.

Structural - parametric modeling was carried out and structural - parametric models of separation processes were developed. A situational model of the separation process of bulk materials has been developed.

The main indicators that have the greatest impact on the quality of the final product after the separation process have been determined and substantiated.

A mathematical model for assessing the quality of separation has been developed and the main indicators have been selected that have the greatest impact on the efficiency of the process, optimization of technological modes of the production process at the modern level.

The research carried out and the results obtained make it possible to move on to the development of effective automated software and hardware systems for controlling the separation of bulk materials using intelligent technologies.

References

[1] Blagoveshchenskaya M. M., Zlobin L. A. 2005. *Information technology of process control systems*. Izdatel'stvo Vysshaya shkola [High school publishing house], 768 p.

[2] Blagoveshchenskaya M.M. (2009). *Basics of stabilization of the preparation of multicomponent masses.* Izdatel'stvo Frantera [Frantera publishing house]. 281 p.
[3] Balykhin M.G., Blagoveshchensky I.G., Nazoykin E.A., Blagoveshchensky V.G. (2019). Adaptive control system with identifier of non-stationary technological processes in the food industry. In the conference proceedings: Intelligent systems and technologies in the food industry. Izdatel'skiy kompleks MGUPP [*Publishing complex MGUPP*], 32 – 39.
[4] Balykhin M.G., Borzov A.B., Blagoveshchensky I.G. (2017). The architecture and basic concept of creating an intelligent expert system of food quality control. *Pishchevaya promyshlennost' [Food Industry]*, 11, 60 - 63.
[5] Balykhin M.G., Borzov A.B., Blagoveshchensky I.G. (2017). *Methodological foundations of creating expert systems for monitoring and forecasting the quality of food products using intelligent technologies*. Izdatel'stvo Frantera [Frantera publishing house. 395 p.
[6] Karelina E.B., Blagoveshchenskaya M.M., Blagoveshchenskiy V.G., Kleho D.Yu., Blagoveshchenskiy I.G. 2019. Integration of adaptive control into the technological processes of the food industry. In the collection: Intelligent systems and technologies in the food industry. 81-89pp.
[7] Blagoveshchenskaya M.M., Blagoveshchensky I.G., Nazoykin E.A. (2015). The technique of automatic assessment of the quality of food products based on the theory of artificial neural networks. *Pishchevaya promyshlennost' [Food Industry]*, 2, 42 - 45.
[8] Blagoveshchenskaya M.M., Santon Kunnikhan M.P. (2017). The structure of dosing control systems using neural networks. *In the collection of conference materials: University Student Conference of Students and Young Scientists "Science Day"* (pp. 263 – 267). Izdatel'skiy kompleks MGUPP [Publishing complex MGUPP].
[9] Blagoveshchenskiy I.G., Nosenko S.M. (2015). Expert intellectual system for monitoring the process of molding fondant sweets using the vision system. *Food Industry, [Pishchevaya promyshlennost']*, 6, 32 - 36.
[10] Garev K.V., Blagoveshchenskiy I.G., Nazoykin E.A., Blagoveshchensky V.G., Makarovskaya Z.V. (2019). Using technical vision as an innovative solution in smart home systems. In the collection: Intelligent systems and technologies in the food industry industries Conference proceedings (pp. 47 – 52). Publishing complex MGUPP [*Publishing complex MGUPP].*
[11] Ivanov Y.V., Blagoveshchenskaya M.M., Blagoveshchenskiy I.G. (2012). Automation of the process of molding candy masses using a digital video camera (CVC) sensor. *Materials of the first international scientific and practical conference - the exhibition "Planning and ensuring the training and retraining of personnel for the food industry and medicine."* (pp. 215 – 218). Publishing complex MGUPP [Publishing complex MGUPP]
[12] Krylova L.A., Blagoveshchenskiy V.G., Tatarinov A.V. (2017). Development of intelligent hardware and software systems for monitoring the separation of dispersed food masses based on intelligent technologies. *In the book: The Development of the Food and Processing Industry of Russia: Personnel and Science* (pp. 199 – 201). Izdatel'skiy kompleks MGUPP [Publishing complex MGUPP].
[13] Nazoykin E.A., Blagoveshchenskiy I.G., Sincha V.M., Zhirov M.V., Mitin V.V. (2019). The use of simulation to identify the state of enterprises in the food industry. *In the collection: Intelligent systems and technologies in the food industry industries Conference proceedings* (pp. 147 – 155). Izdatel'skiy kompleks MGUPP [Publishing complex MGUPP].
[14] Nosenko S.M., Blagoveshchenskiy I.G., Shaverin A.V., Blagoveshchenskaya M.M. (2014). Automation of control of taste indicators of chocolate products using intelligent technologies // *Confectionery and bakery production [Konditerskoye i khlebopekarnoye proizvodstvo]*, 10 (153), 56 - 59.
[15] Petryakov A.N., Blagoveshchenskaya M.M., Blagoveshchensky V.G., Krylova L.A. (2018). The use of object-oriented programming methods to control the quality indicators of confectionery products. *Zhurnal "Konditerskiye izdeliya i khlebopecheniye" [Confectionery and Bakery Magazine],* no. 5-6 (176), 21-23.
[16] Petryakov A.N., Blagoveshchenskaya M.M., Blagoveshchenskiy V.G., Mitin V.V., Blagoveshchenskiy I.G. (2019). Improving the quality of identification and positioning of an object in digital stereo images using depth mapping algorithms. *In the collection: Intelligent systems and technologies in the food industry industries. Conference proceedings* (pp. 133 – 138). Publishing complex MGUPP [Publishing complex MGUPP].
[17] Savostin S.D., Blagoveshchenskaya M.M., Blagoveshchenskiy I.G. (2016). *Automation of control of flour quality indicators during grinding using intelligent technologies.* Frantera. [Frantera publishing house].

[18] Shaverin A.V., Blagoveshchenskaya M.M., Blagoveshchenskiy I.G. (2012). Automation of the control of organoleptic quality indicators of chocolate products. *Materials of the first international scientific-practical conference - the exhibition "Planning and providing training and retraining of personnel for the food industry and medicine."* (pp. 209 – 212). Publishing complex MGUPP [Publishing complex MGUPP].
[19] Balykhin M.G., Blagoveshchenskaya M.M., Blagoveshchenskiy I.G., Nazoykin E.A., Naumov R.R. (2018). Analysis of the functioning of the production processes of jelly sweets by the method of simulation. *Pishchevaya promyshlennost' [Food industry*]. 11, 56-59.
[20] Balykhin M.G., Borzov A.B., Blagoveshchensky I.G. (2017). The architecture and basic concept of creating an intelligent expert system of food quality control. *Pishchevaya promyshlennost' [Food Industry],* 11, 60 - 63.
[21] Blagoveshchenskaya M.M., Sulimov V.D., Shkapov P.M. (2010). A development methodology for modeling and diagnosing hydromechanical systems of food production according to their dynamic characteristics // High Intellectual Technologies and Innovations in Education and Science: Materials of the XVII International Scientific-Method. conf. February 11-12, 2010, Izdatel'stvo SPbGPU *[Publishing House SPbSPU],* 95-98.
[22] Blagoveshchenskiy I.G. (2018). *Methodological foundations of creating expert systems for monitoring and forecasting the quality of food products using intellectual technologies*: Abstract. dis .... doctor. tech. Sciences: 05.13.06. - M., 2018, - 46.
[23] Bychkov I.A., Blagoveshchenskaya M.M., Nosenko A.S., Blagoveshchensky I.G. (2015). The method of generalized interval estimates to support group expert decisions in the face of uncertainty. Khraneniye i pererabotka sel'khozsyr'ya [Storage and processing of agricultural raw materials], 4, 63-65.
[24] Ivanov Y.V., Blagoveshchenskaya M.M., Blagoveshchenskiy I.G. (2012). Automation of the process of molding candy masses using a digital video camera (CVC) sensor. *Materials of the first international scientific and practical conference - the exhibition "Planning and ensuring the training and retraining of personnel for the food industry and medicine."* (pp. 215 – 218). Publishing complex MGUPP [Publishing complex MGUPP].
[25] Karelina E.B., Blagoveshchenskaya M.M., Blagoveshchensky V.G., Klekho D.Yu., Blagoveshchensky I.G. (2019). Integration of adaptive management in the technological processes of the food industry. In the conference proceedings: Intelligent systems and technologies in the food industry, Izdatel'stvo MGUPP *[Publishing house MGUPP],* 81-89.
[26] Mokrushin S.A., Blagoveshchenskiy I.G., Blagoveshchenskaya M.M., Zhirov M.V., Chuvakhin S.V. (2019). Creation, adjustment and testing of a system for automatic control of the technological process of sterilization of canned food using a simulation model. In the conference proceedings Intelligent Systems and Technologies in Food Industries. Izdatel'stvo MGUPP *[Publishing house MGUPP]*, 100-109.
[27] Mokrushin S.A., Blagoveshchensky I.G., Nazoykin E.A., Mitin V.V., Chuvakhin S.V. (2019). A simulation model of the canned food sterilization process as an effective tool in the development and modernization of sterilization plants. In the collection of conference materials Intelligent systems and technologies in the food industry. Izdatel'stvo MGUPP *[Publishing house MGUPP]*, 90-99.
[28] Nazoykin E.A., Blagoveshchenskiy I.G. (2018). Identification of production processes using simulation methods. In the collection of conference materials with international participation. Current status and prospects for the development of packaging in the food industry. *Izdatel'stvo MGUPP [Publishing house MGUPP*], 97-102.
[29] Nazoykin E.A., Blagoveshchensky I.G. (2019). The use of simulation methods to identify the processes of dough preparation at the bakery. *In the collection: Ninth All-Russian Scientific and Practical Conference on Simulation Modeling and its Application in Science and Industry Conference proceedings.* 468-472.
[30] Nazoykin E.A., Blagoveshchensky I.G., Balykhin M.G., Makarovskaya Z.V., Sumerin V.A. (2019). Simulation of the test preparation department at the bakery with ANYLOGIC. In the collection of materials of the conference Intelligent systems and technologies in the food industry. Izdatel'stvo MGUPP *[Publishing house MGUPP],* 139-146.
[31] Nazoykin E.A., Blagoveshchensky I.G., Blagoveshchenskaya M.M., Naumov R.R. (2019). Identification of the processes of production of marmalade masses using simulation modeling. Pishchevaya industriya [Food Industry], 1 (39), 40-41.
[32] Petryakov A.N., Blagoveshchenskaya M.M., Blagoveshchenskiy V.G., Mitin V.V., Blagoveshchenskiy I.G. (2019). Improving the quality of identification and positioning of an object in

digital stereo images using algorithms for constructing a depth map. In the conference proceedings: Intelligent systems and technologies in the food industry. Izdatel'skiy kompleks MGUPP *[Publishing complex MGUPP]*, 133-138.

[33] Petryakov A.N., Blagoveshchenskaya M.M., Nosenko A.S., Blagoveshchenskiy I.G. (2015).The use of genetic algorithms and particle swarm methods in the production of animal feed. Pishchevaya promyshlennost' *[Food industry]*. 6, 18-21.

[34] Blagoveshchenskiy I.G. (2018). *Methodological foundations of creating expert systems for monitoring and forecasting the quality of food products using intellectual technologies*: Abstract. dis .... doctor. tech. Sciences: 05.13.06. - M., 2018, - 46.

[35] Blagoveshchensky V.G., Krylova L.A., Maksimov A.S. (2017). Development of a hardware and software complex for monitoring halva production. In the book: Development of the food and processing industry in Russia: personnel and science. Izdatel'skiy kompleks MGUPP [*Publishing complex MGUPP*], 196 -199.

[36] Blagoveshchensky V.G., Krylova L.A., Nikitushkina M.Yu. (2018). Automation of the sugar syrup preparation process. In the collection: Advanced food technologies: state, trends, points of growth. Collection of scientific papers of the I scientific-practical conference with international participation. Izdatel'skiy kompleks MGUPP [*Publishing complex MGUPP*], 663-667.

[37] Blagoveshchensky V.G., Nikitushkina M.Yu. (2017). Automation of the fondant syrup preparation process. In the book: Development of the food and processing industry in Russia: personnel and science. Izdatel'skiy kompleks MGUPP [*Publishing complex MGUPP*], 202 -205.

[38] Naumov R.R., Nazojjkin E.A., Blagoveshchenskii I.G., Blagoveshchenskaya M.M. (2018). Application of agent technologies in analysis of production processes of food production. V sbornike: Peredovye pishchevye tekhnologii: sostoyanie, trendy, tochki rosta Sbornik nauchnyh trudov I nauchno-prakticheskoj konferencii s mezhdunarodnym uchastiem. 711-715.

[39] Nikitushkina M.Yu., Krylova L.A., Blagoveshchensky V.G. (2017) Development of an expert quality control system in the process of halva preparation. *University-wide student conference of students and young scientists "Science Day". Collection of conference materials,* 294-301.

[40] Petrov A.Yu., Blagoveshchenskaya M.M., Blagoveshchenskiy V.G., Ionov A.V., Blagoveshchenskiy I.G. (2019). The main principles in building a computer vision system in the baking industry. In the conference proceedings: Intelligent systems and technologies in the food industry. Izdatel'skiy kompleks MGUPP *[Publishing complex MGUPP]*, 121-126.

Modeling of the thermal evolution of the cores of icy giant satellites

E V Kronrod, V A Kronrod, O L Kuskov

Vernadsky Institute of Geochemistry and Analytical Chemistry, Russian Academy of Sciences (RAS), Kosygina St., 19, Moscow, 119991 Russia

e.kronrod@gmail.com

Abstract. Questions on the composition, aggregate state, size and physical properties of the giant planets satellites cores, their thermal evolution, as well as the mechanisms of their formation, are still the subject of numerous discussions. Models of the internal structure of large icy satellites impose restrictions on the composition of satellite cores in accordance with the matter of L/LL or CI chondrites or their mixture. The temperature distributions in the cores of satellites largely determine the measure of hydration of their silicate component, the presence or absence of internal metal cores. In this paper, we present the results of numerical modeling of nonstationary temperature regimes in cores with allowance for convective transfer processes. According to the results of calculations, it can be concluded that in the presence of convection, the composition of the core largely determines the thermal evolution of the core. For example, the temperature difference between silicate and hydrosilicate composition, can reach ≈ 600 K.

1. Introduction

As a result of research missions to Jupiter and Saturn implemented in recent decades ("Galileo", "Cassini–Huygens"), a fundamentally new information about the ice giant satellites Ganymede, Callisto and Titan was obtained, which made it possible to construct models of the internal structure of the satellites. The main existing models of satellites include a water-ice shell (± undifferentiated rock-ice mantle), an iron silicate core (± Fe–FeS inner core) [1, 2]. Questions on the composition, state of aggregation, size and physical properties of satellite cores, their thermal evolution, as well as the mechanisms of its formation are still the subject of numerous discussions. Models of the internal structure of large ice satellites [1, 2, 3] impose restrictions on the composition of satellite cores in accordance with the matter of ordinary (L / LL) or carbonaceous (CI) chondrites or their mixture. Temperature distribution in cores largely determine the degree of hydration of their silicate component, the presence or absence of internal metal core. In this paper, we present the results on estimates of non-stationary temperature regimes in cores taking into account the processes of convective heat transfer and the composition of the core.

2. Model Description

At the final stage of accretion from rocky-ice planetesimals, the temperature in the near-surface region of the satellite increases and reaches the melting point of ice. As a result, the separation of the stony-iron component, its migration to the center of the satellite and the formation of the silicate-iron core occurs. The process of heating of the proto-core by the energy of radioactive decay to a temperature exceeding the melting point of ice (~500 K), redistribution of ice, water and rock material with the formation of a homogeneous core occurs over a time of about 500 million years [4]. Subsequently, radioactive sources of energy become the cause of the thermal evolution of the core.

It is assumed that satellite core consists of a mixture of L / LL and CI - chondritic substance, further with indices Si (silicate), HS (hydrosilicate), respectively. The mass concentrations of CI chondrites (C_{HS}) and L / LL chondrites ($C_{Sil} = 1 - C_{HS}$) are specified. The cores of icy satellites, after formation, are heated as a consequence of the radioactive decay of radioactive isotopes. The approximation of the radiogenic heat release of chondrite matter by the radioactive decay equation in the interval of 500–4.500 Ma from CAI leads to the initial value of the power of radiogenic sources $H_{0\,Si} = 3.1334\cdot10^{-11}$ W/kg and $H_{0\,HS} = 2.1336\cdot10^{-11}$ W/kg, and also allows determine the constants of radioactive decay of L/LL-chondrite and CI-chondrite matter: $\lambda_{Si} = -1.4457\cdot10^{-17}$, $\lambda_{HS} = -1.4553\cdot10^{-17}$.

The temperature distributions in the cores are found as a result of the numerical solution of a one-dimensional nonstationary heat equation with initial (500 K) and boundary (360 K) conditions.

Two mechanisms of heat transfer are considered: conductive (diffuse mechanism) and convective (transfer of liquid volumes with different temperatures). The transition from conductive to convective regime is estimated by Rayleigh number (Ra) for the case of internal uniformly distributed in the volume energy sources [5, 6]:

$$Ra(t)= \rho^2{}_m\, \alpha_m g_m\, R_0^5 Q(t)/(k_m \kappa_m \eta_m) \qquad (1)$$

where the subscript m denotes the average parameters in the inner convective zone with the radius R_0, g is gravitational acceleration; $R_0 = R_C - \delta$, R_C – the radius of the core, δ – the thickness of the thermal boundary layer (the layer without convection). The thickness of the thermal boundary layer δ is determined from the condition on the critical Rayleigh number $Ra_{cr} \approx 100$ [5].

Heat transfer in the convective zone is modeled by the effective thermal conductivity coefficient: $K_{ef} = Nu\ K_{conduct}$, [6]. Here Nu is the dimensionless Nusselt number; $Nu = 1.04\ (Ra / Ra_{crit})^{1/3}$, $Ra_{crit} \approx 1000$ [7]. In the case when $Ra < Ra_{crit}$, $K_{ef} = K_{conduct}$.

The temperature distributions in the core are found as a result of the numerical solution of a one-dimensional nonstationary equation that takes into account both conductive and convective transfers using the effective transfer coefficient K_{ef}, as well as the processes of heating of the core material due to the radiation energy:

$$\rho C_p \frac{\partial T}{\partial t} = \frac{1}{r^2}\frac{\partial}{\partial r}\left[r^2 K_{ef}\frac{\partial T}{\partial r}\right] + Aexp(-\lambda t) \qquad (2)$$

where r is the current radius, ρ – the density, C – the heat capacity, t – the time, A – the heat release per unit mass of the core at the initial moment of time, λ – decay constant.

The initial conditions are: $t = 500$ million years, $T = 500$ K, boundary conditions: $R = R_{core}$, $T = 360$ K; $R = 0$, $\partial T / \partial r = 0$.

Equation (2) is solved by a finite-difference method. The accuracy of the calculations was checked using an analytical solution.

In the calculations, the following values were set for silicates and hydrosilicates, respectively: ρ = 3620 and 2800 kg m^{-3}; heat capacity C_p = 920 and 1360 J / kg · K; thermal conductivity K - 4.2 W / m K and 2.95 W / m K. The viscosity of silicates–hydrosilicates mixture is calculated using the isostress model [8]:

$$\eta_{HS} = 4\cdot10^{19\cdot fG} \cdot 4.9\cdot10^{8\cdot fS} \cdot exp(f_S\cdot 23\cdot(\ T_{0mlt} /T)), \qquad (3)$$

where f_{HS}, f_S are volume fractions of hydrosilicates, silicates; T, T_{0mlt}, is the temperature in the computational domain, the melting point of silicates corresponding to P–T conditions. The viscosity of the composite in (3) is calculated from the average physical parameters of the mixture in the calculated volume.

The viscosity of silicates is determined depending on temperature and pressure similarly to [9], the viscosity of hydrosilicates is assumed to be constant $4{\cdot}10^{19}$ Pa s in the entire range of temperatures and pressures [10]. Possible phase transitions of hydrosilicates to silicates in the temperature range 900 K are not considered.

3. Results

Calculations of thermal evolution in cores of giant satellites with hydrosilicate concentrations from 0 to 100% and radii R_{core} from 500 to 2000 km were performed.

During the heating of the core by radiogenic sources uniformly distributed over the volume, the temperature in the core gradually increases, and as a consequence, the viscosity of the substance decreases, which leads to the formation of a region of possible convection. Composition has a significant effect on thermal evolution. The higher the content of hydrosilicates in the core, the faster convection occurs. For pure hydrosilicates ($C_{Sil} = 0$), convection starts immediately, the convection zone occupies almost the entire core region, and viscosity decreases during cooling, which leads to a decrease in the convective zone radius. The convection intensity depends on the Rayleigh number Ra (Figure 1a) - a dimensionless parameter that determines the behavior of a material under the influence of a temperature gradient. The results are shown for core radius R_{core} = 1500 km and various concentrations of silicates. Intense convection is observed at times of 1,5-3,5 Byr from CAI.

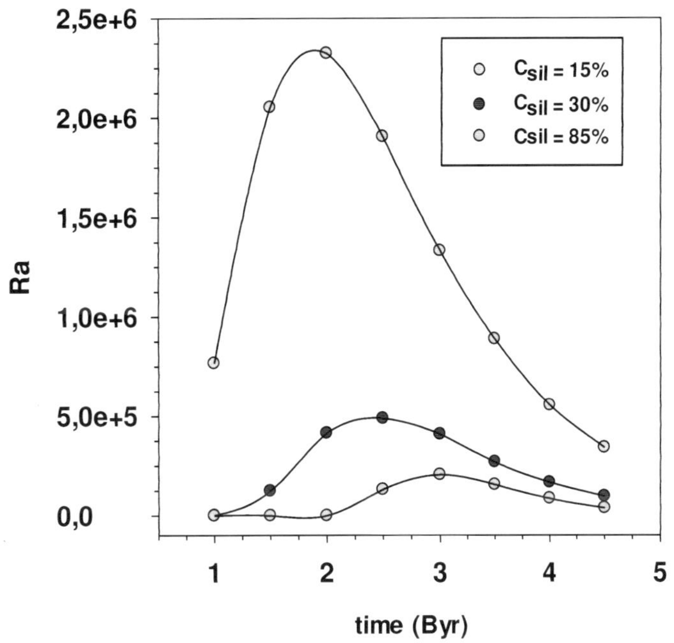

Figure 1. Dependence of the Rayleigh number Ra from the time in the interval 500 – 4.500 Ma for various content of silicates (C_{Sil}). The core radius R_{core} = 1500 km.

Figure 2 shows the time dependence of K (the coefficient of nonstationarity of heating) (a) and surface heat flux Q_s (b) for core radius R_{core} = 1500 km. With an increase in the content of silicates in the core, the surface heat flux increases (Figure 2a). Coefficient K (Figure 2a) is ~1 only at times close to the present (4.5 billion years from CAI). This fact should be taken into account for calculating thermal models of satellites when considering stationary models of thermal conductivity.

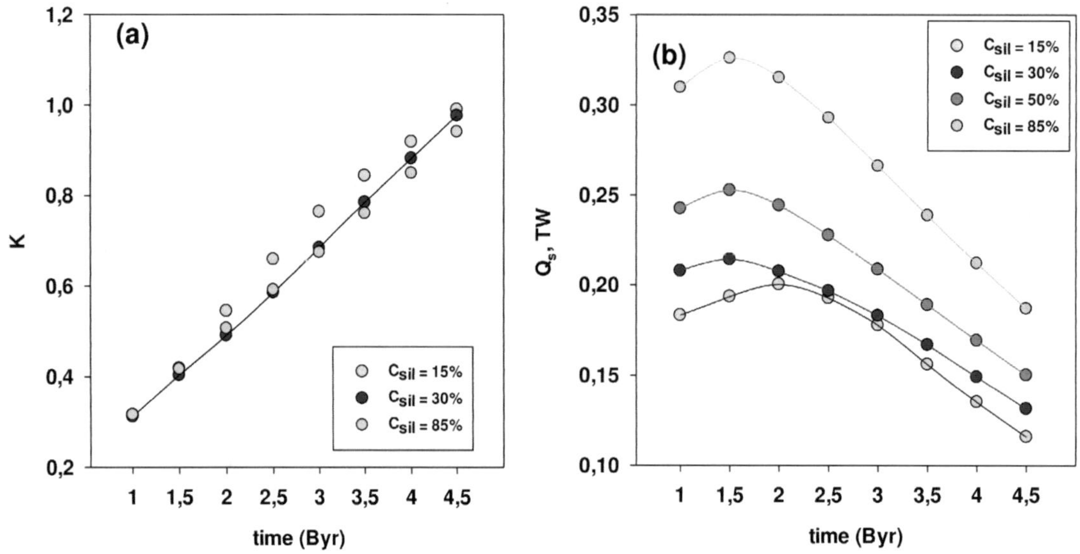

Figure 2. Time dependence of K (the coefficient of nonstationarity of heating) (a) and surface heat flux Q_s (b). R_{core} = 1500 km.

Figure 3 and Figure 4 show the results of calculations of the temperature distribution T as a function of radius and time for R_{core} = 1500 km. At a low content of silicates (15%, Fig.3b), the temperature in the core increases from the initial moment (500 Ma from CAI, 500 K in the center), reaching the highest value to 2-2.5 Byr (~ 900 K in the center) and then gradually decreases to ~ 800 K in the center at the present stage (4.5 Byr). In this case, the temperature of 900 K (T of the phase transition) is not reached at any time. The transition from the central convective zone to the outer conductive zone is clearly visible at a radius of ~ 1200-1300 km (Figure 3b), which is seen in the graph as a change in the slope of the temperature curves. With a large amount of silicates (85%, fig. 1a), the temperature profile shifts towards higher values, reaching a maximum by 3.5 Byr (1525 K), then the temperature begins to decrease (~ 1500 K at present time). Figure 4 shows the current temperature distribution (4.5 billion years from CAI). For a satellite with a radius of 1500 km, the spread of possible temperatures in the center is from 420 K (at Csil = 0) to temperatures of ~ 1600 K, which is close to the melting point of iron (at C_{sil} = 100%).

Temperature ~ 900 K marks the possibility of the beginning of the dehydration process – a phase transition from hydrosilicates to silicates with the release of water. Figure 5 shows the largest radius at which a phase transition can occur at different radii of R_{core} and the content of silicates (Sil) in the core. At $R_{core} \leq$ 1500 km, phase dehydration is impossible for silicate content < 30%.

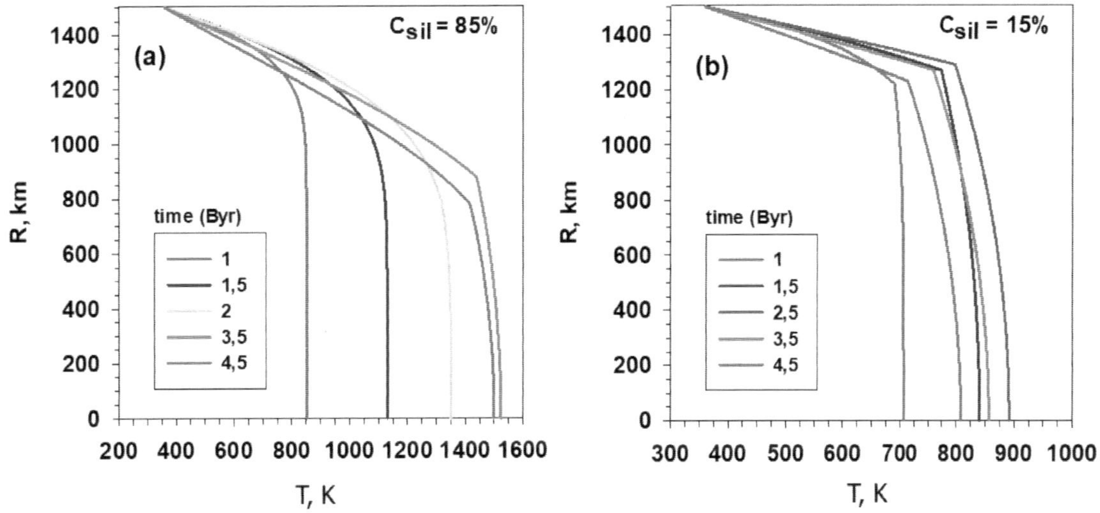

Figure 3. Temperature distributions (T) in the core (Rcore = 1500 km) as a function of radius (R) and time with a silicate content of 85% (a) and 15% (b)

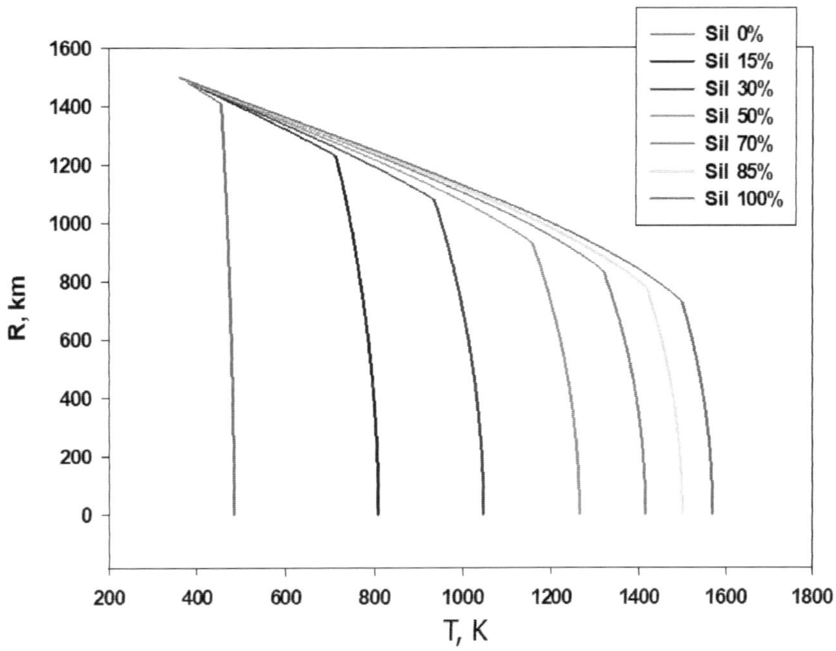

Figure 4. Temperature distribution for a satellite with a core radius of Rcore = 1500 km at present time (t = 4.5 Byr from CAI)

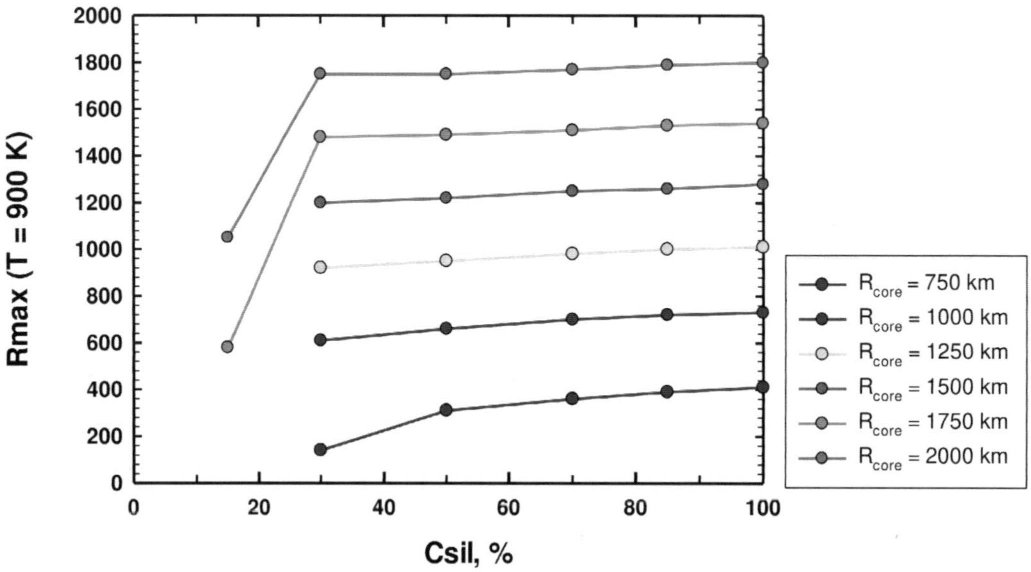

Figure 5. The maximum radius at which a temperature of 900 K is reached, depending on the concentration of silicates in the satellite core. The color indicates the outer radii of the core.

4. Conclusions

Composition has a significant effect on thermal evolution. With a high content of hydrosilicates, the phase transition temperature is not reached. The transition to convection takes place by 2 billion years and can continue until the present time with a high content of hydrosilicates. Stationary models give an error in the temperature estimates. Based on the data obtained, it is possible to build models of the internal structure of satellites.

Acknowledgments The work was performed as part of the state assignment of Vernadsky Institute of Geochemistry and Analytical Chemistry № 0137-2019-0004.

References

[1] Fortes A.D. Titan's internal structure and the evolutionary consequences // Planetary and Space Science. 2012. V. 60. P. 10-17. doi 10.1016/j.pss.2011.04.010.
[2] Dunaeva A.N., Kronrod V.A., & Kuskov O.L. Physico-chemical models of the internal structure of partially differentiated Titan.// Geochemistry International, 2016. 54(1), P. 27-47.
[3] Kronrod V.A., Kuskov O.L. Chemical differentiation of the Galilean satellites of Jupiter: 4. Isochemical models for the compositions of Io, Europa, and Ganymede // Geochem. Int. 2006. V. 44. P. 529-546. doi 10.1134/S0016702906060012.
[4] Castillo-Rogez J.C., Lunine J.I. Evolution of Titan's rocky core constrained by Cassini observations // Geophys. Res. Lett. 2010. V. 37. L20205.
[5] Grasset O., Sotin C., Deschamps F. On the internal structure and dynamics of Titan // Planet. Space Sci. 2000. V. 48. P. 617–636. doi 10.1016/S0032-0633(00)00039-8.
[6] Czechowski L., Witek P. Comparison of early evolutions of mimas and enceladus // Acta Geophysica, 2015. V. 63(3). P. 900-921.
[7] Turcotte, D.L., and G. Schubert (2002), Geodynamics, 2nd ed., Cambridge University Press, Cambridge, 465 pp.
[8] Ji S. and Zhao P. 1993. Flow laws of multiphase rocks calculated from experimental data on the

constituent phases// Earth and Planetary Science Letters, 117, 181-187
[9] Kimura J., Nakagawa T., Kurita K. 2009. Size and compositional constraints of Ganymede's metallic core for driving an active dynamo. Icarus 202. 216–224.
[10] Hilairet N., Reynard B., Wang Y., Daniel I., Merkel S., Nishiyama N., Petitgirard S. 2007. High-Pressure Creep of Serpentine, Interseismic Deformation, and Initiation of Subduction // SCIENCE. V. 318. 1910-1913.

IOP Conf. Series: Materials Science and Engineering 1191 (2021) 012020 doi:10.1088/1757-899X/1191/1/012020

Analysis of residual stress and strain in the cylindrical part of the composite pressure vessel with load-sharing metallic liner

Sarbayev B.S.[1], Zhang X.[1]

[1]Bauman Moscow State Technical University, 2-th Bauman str. Moscow 105005, Russia

E-mail: bssarbayev@mail.ru; zxbmstu@qq.com

Abstract. As a result of experimental studies of the composite pressure vessel with load-shearing metallic liner during the unloading and cyclic-loading processes, the deformation of structure is nonlinear [1,2]. A method for purpose of describing and analyzing residual stress and strain in the cylindrical part of a composite pressure vessel type 3 after autofrettage is developed, which is based on the linear-elastic model of the fibrous composites and the relations of the theory of plasticity with translational hardening for isotropic material [3]. The theoretical results obtained from constitutive equations are written in matrix form and fit the experimental data well. In comparison, the results based on finite element method and classical Prandtl-Reuss flow rule are proposed. The results show that it is necessary to consider the plastic deformation in metallic liner of composite pressure vessel type 3 when analyzing the stress-strain state.

1. Introduction

Composite pressure vessels are widely used in practically all the fields of modern technology. One of the most common structures is a composite pressure vessel with load-sharing metallic liner whose domes have helical-wrapped reinforcements, and its cylindrical part is reinforced with both circumferential and helical layers of fibrous composite material (pressure vessel type 3 [4]). Elastic-plastic strains, which occur in liner material when the vessel is subjected to intense force loading, leading to the formation of residual stresses and strain, hysteresis loops under low-cycle loading. To ensure reliable and safe operation of the vessel and reduce the risk of emergencies, it is necessary to adequately evaluate the stress-strain state in both the liner and the multilayer composite material.

Consideration of plastic strains is also necessary when justifying the operational modes of some technological operations. Particularly, autofrettage should be taken during manufacturing the preliminary vessel which increases the operating pressure. After the operation irreversible plastic deformations occur in the liner material. The yield stress of liner material is raised due to the hardening effect, and linear elastic deformation takes place during subsequent cyclic loading liner. However, during the unloading process the liner is compressed by external contact pressure at same time, and it will lead to an undesirable buckling of the thin-walled liner.

The relations based on the deformation theory of plasticity [1,2,5] or Prandtl-Reuss flow rule with isotropic hardening [6,7] are mostly used in many publications. The application of these mentioned approaches does not allow to adequately describe the unloading and repeated loading of pressure vessel type 3. In particular, the indicated models of metal deformation are not able to represent the hysteresis loops which were observed in experiments [2]. On the basis of constitutive equations of theory of plasticity with translational hardening for isotropic solid [3], a method for calculating residual stresses and strains in the cylindrical part of a composite pressure vessel type 3 is proposed in previous work. In this case the theoretical relationships are given which can be used for obtaining the permissible pressure for autofrettage and the bearing capacity of this type of pressure vessels. The calculation results derived with the aid of finite element method and Prandtl-Reuss relations are presented for comparison. A linear elastic model for the unidirectional fibrous composite material is used for both cases.

2. Elastic model.

Consider a metallic-composite pressure vessel of type 3 with radius of cylindrical part R_m. The vessel is loaded with internal pressure *P*. The cylindrical shell consists of a multilayer material. Excepting the metallic liner with thickness $h_м$, the multilayer wall of shell contains a laminated composite fibrous ply consisting of circumferential layers with thickness $h_к$ and symmetrical angle-ply helical layers with thickness h_c.

Because the section of cylindrical part is far enough away from the section, where combined the cylindrical part and dome, the bending stresses are zero. We separate out an infinitely small element from the cylindrical part as shown in Fig 1. Suppose that the axis *OX* coincides with the meridian of cylindrical part, axis *OY* – shear to its circular section, axis *OZ* – normal to the surface of cylindrical part. In this case, the equilibrium equations for the thin-walled cylindrical part have the following forms

$$\sigma_{11\text{м}} h_\text{м} + \sigma_{11\text{км}} h_\text{км} = \frac{1}{2} PR; \quad \sigma_{22\text{м}} h_\text{м} + \sigma_{22\text{км}} h_\text{км} = PR. \tag{1}$$

in which $h_\text{км} = h_\text{к} + h_\text{c}$. $H = h_\text{м} + h_\text{км}$ corresponds the total thickness of cylindrical part.

The stress-strain state of cylindrical part of is described by the vectors of mean stresses $\{\sigma_\text{c}\}$ and mean strains $\{\varepsilon_\text{c}\}$. They are related through Hooke's law in the range of elastic limit and take the form

$$\{\sigma_c\} = [G_c]\{\varepsilon_c\}, \tag{2}$$

where $\{\sigma_c\} = (\sigma_{11c}, \quad \sigma_{22c}, \quad \sigma_{12c})^\text{T}$, $\{\varepsilon_c\} = (\varepsilon_{11c}, \quad \varepsilon_{22c}, \quad \gamma_{12c})^\text{T}$.

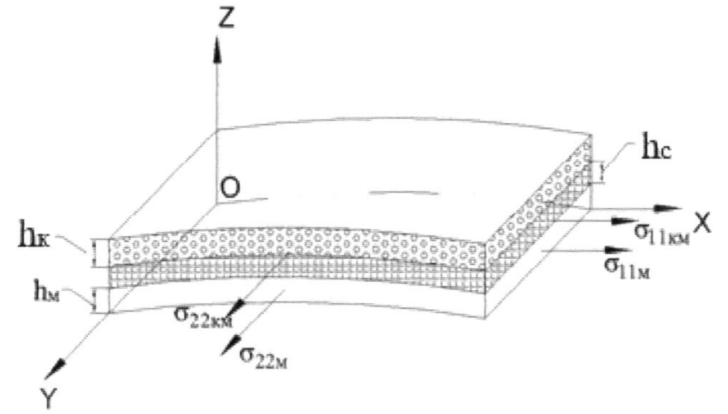

Figure. 1. Stresses acting in the layers of the cylindrical part of a type 3 vessel.

Write the stiffness matrix of multilayer material in cylindrical part as

$$[G_c] = \begin{bmatrix} g_{xx} & g_{xy} & 0 \\ g_{xy} & g_{yy} & 0 \\ 0 & 0 & g_{tt} \end{bmatrix}.$$

Using the following relation, we can obtain the components in this matrix

$$[G_c] = [D_\text{м}]\delta_\text{м} + [D_\text{км}]\delta_\text{км}. \tag{3}$$

Here $[D_\text{м}]$ – elastic stiffness matrix of metallic liner; $\delta_\text{м} = \frac{h_\text{м}}{H}$, $\delta_\text{км} = 1 - \delta_\text{м}$. For matrix $[D_\text{м}]$ follows equality

$$[D_{\text{м}}]=\begin{bmatrix}\frac{E}{1-\nu^2} & \frac{E}{1-\nu^2}\nu & 0\\ \frac{E}{1-\nu^2}\nu & \frac{E}{1-\nu^2} & 0\\ 0 & 0 & \frac{E}{2(1+\nu)}\end{bmatrix}.$$

Here E – Young's modulus of liner material, ν – Poisson's ratio of liner material.

The stiffness matrix of multilayer composite solid can be found with the aid of

$$[D_{\text{км}}]=\sum_{i=1}^{k}[T_{1(i)}][D_{\text{к}}][T_{1(i)}]^{T}\delta_{(i)},$$

where $[D_{\text{к}}]$ – stiffness matrix of monolayer composite material, it takes following form

$$[D_{\text{к}}]=\begin{bmatrix}\frac{E_1}{1-\nu_{12}\nu_{21}} & \frac{E_1\nu_{21}}{1-\nu_{12}\nu_{21}} & 0\\ \frac{E_2\nu_{12}}{1-\nu_{12}\nu_{21}} & \frac{E_2}{1-\nu_{12}\nu_{21}} & 0\\ 0 & 0 & G_{12}\end{bmatrix}.$$

Here E_1, E_2, G_{12} – elastic modulus of monolayer, ν_{12}, ν_{21} – Poisson's ratio of monolayer. Express the transformation matrix for rotation of the coordinate as

$$[T_{1i}]=\begin{pmatrix}\cos^2\varphi_i & \sin^2\varphi_i & \sin 2\varphi_i\\ \sin^2\varphi_i & \cos^2\varphi_i & -\sin 2\varphi_i\\ -\sin\varphi_i\cos\varphi_i & \sin\varphi_i\cos\varphi_i & \cos 2\varphi_i\end{pmatrix},$$

where φ_i $(i=1,2,\dots,k)$ – winding angle.

From formula (2), we get

$$\{\varepsilon_c\}=[G_c]^{-1}\{\sigma_c\}. \tag{4}$$

Matrix $[G_c]^{-1}$ can be found from the following equation

$$[G_c]^{-1}=\begin{bmatrix}\frac{g_{yy}}{\Delta} & \frac{-g_{xy}}{\Delta} & 0\\ \frac{-g_{xy}}{\Delta} & \frac{g_{xx}}{\Delta} & 0\\ 0 & 0 & \frac{1}{g_{tt}}\end{bmatrix},$$

where $\Delta=g_{xx}g_{yy}-g_{xy}{}^2$. Then from formula:

$$\{\varepsilon_{\text{к}i}\}=[T_{2i}]^{-1}\{\varepsilon_c\},$$

and $[T_{2i}]^{-1}=[T_{1i}]^{\text{T}}$ as well, we can get the strains in i - th monolayer. The equation for the stresses in composite monolayer takes the form

$$\{\sigma_{\text{к}i}\}=[D_{\text{к}}]\{\varepsilon_{\text{к}i}\}.$$

Strains of liner material are equal to the mean strains of multilayer solid i.e., $\{\varepsilon_{\text{м}}\}=\{\varepsilon_c\}$, From here for the stresses in liner material, we get

$$\{\sigma_{\text{м}}\}=[D_{\text{м}}]\{\varepsilon_c\}. \tag{5}$$

In this case we can present the mean stresses in the following forms

$$\sigma_{11c}=\frac{PR}{2H},\quad \sigma_{22c}=\frac{PR}{H}. \tag{6}$$

Meanwhile, in accordance with equation (4) we arrive at $\gamma_{12c}=0$ i.e., the shear strain in cylindrical part does not exist.

The pressure P_y, which induces the plastic strain in the metallic layer can be found taking $\sigma_i=\sigma_y$ in which σ_y is the yield stress of the liner material. In this case, stress intensity in liner material follows as

$$\sigma_i=\sqrt{\sigma_{11\text{м}}{}^2-\sigma_{11\text{м}}\sigma_{22\text{м}}+\sigma_{22\text{м}}{}^2}. \tag{7}$$

Then we can write the expression for calculating the pressure P_y while the plastic flow begins in liner. After transformation we arrive at

$$P_y=\frac{2H}{R}\frac{\sigma_y}{\sqrt{b_1^2-b_1b_2+b_2^2}}, \tag{8}$$

where $b_1 = E\frac{(g_{yy}-2g_{xy})+v(2g_{xx}-g_{xy})}{\Delta(1-v^2)}$, $b_2 = E\frac{v(g_{yy}-2g_{xy})+(2g_{xx}-g_{xy})}{\Delta(1-v^2)}$. The stresses in liner are presented as $\sigma_{11M} = b_1\frac{PR}{2H}$, $\sigma_{22M} = b_2\frac{PR}{2H}$, which correspond the pressure P_y.

2. Plastic model.

In present work a plastic model with translational hardening for isotropic solid [3] is developed in order to describe the plastic deformation of liner material. The corresponding equation of loading surface can be expressed as

$$F(s_{ij}) = \frac{3}{2}(s_{ij} - r_{ij})(s_{ij} - r_{ij}) = \sigma_y^2. \tag{9}$$

In proposed theory the isotropic expansion of loading surface is ignored. A new vector r_{ij} is called back stresses tensor and gives the coordinate of the center of the loading surface in deviator stress space. Meanwhile, to determine the tensor r_{ij}, an integral relationship is given as

$$r_{ij} = \frac{2}{3}R\int_0^{\xi_p} exp\{-\alpha[z_p(\xi_p) - z_p(\xi_p')]\}\frac{de_{ij}^p}{d\xi_p'}d\xi_p'. \tag{10}$$

where $(d\xi_p)^2 = 2/3de_{ij}^p de_{ij}^p$ – Odqvist parameter.

In formula (10) $\alpha > 0, R > 0$ should be fulfilled. These parameters are determined by experimental methods, for example, from the torsion tests of thin-walled tubular sample. A monotonically increasing function z_p is taken as an intrinsic-time parameter and represents the history of loading. Then, we apply expression $f(\xi_p) = 1 + \beta\xi_p$, where $\beta > 0$ – material parameter. The analysis shows that the results specified by the chosen function $f(\xi_p)$ give a good agreement with experimental data.

After the differentiation equation (10), we get the expression of back stress increment

$$dr_{ij} = \frac{2}{3}Rde_{ij}^p - \frac{2}{3}\alpha R\int_0^{\xi_p} exp\{-\alpha[z_p(\xi_p) - z_p(\xi_p')]\}dz_p\frac{de_{ij}^p}{d\xi_p'}d\xi_p'.$$

From here, in view of $dz_p = d\xi_p/f(\xi_p)$ and equation (10), we can obtain

$$dr_{ij} = \frac{2}{3}Rde_{ij}^p - \alpha r_{ij}\frac{d\xi_p}{f(\xi_p)}. \tag{11}$$

According to the associated flow rule, the following relation between deviator stresses s_{ij} and deviator plastic strains e_{ij}^p is true

$$s_{ij} = \frac{2}{3}\sigma_0 de_{ij}^p/d\xi_p + r_{ij}. \tag{12}$$

Then an auxiliary tensor of active stresses $s_{ij}^0 = s_{ij} - r_{ij}$ is introduced for the sake of convenience. Substituting it into equation of yield surface (9), we arrive at $\frac{2}{3}s_{ij}^0 s_{ij}^0 = \sigma_y{}^2$. Since $r_{ij} = 0$, we get Mises criterion from above equation. In conjunction with (12) and $d\sigma_{kk} = 3Kd\varepsilon_{kk}$, plastic deformation in liner material occurs when the conditions $\frac{2}{3}s_{ij}^0 s_{ij}^0 = \sigma_y{}^2$, $s_{ij}^0 ds_{ij} > 0$ $s_{ij}^0 ds_{ij} \leq 0$ are fulfilled; Since $\frac{3}{2}s_{ij}^0 s_{ij}^0 = \sigma_y{}^2$, $s_{ij}^0 ds_{ij} \leq 0$, deformation of liner takes place linear-elastically, i.e., we have $ds_{ij} = 2Gde_{ij}$, $d\sigma_{kk} = 3Kd\varepsilon_{kk}$. Here G, K – shear modulus and bulk modulus respectively.

In numerical analysis the constitutive relations of proposed theory are advisable written in matrix from. Therefore, using equations (9)-(12) and taking following transformations. Applying relation

$$de_{ij}^p = de_{ij} - \frac{ds_{ij}^0 + dr_{ij}}{2G},$$

and formulas (11), (12) as well, after transformations we obtain

$$\left[\left(\frac{3}{2} + \frac{R}{2G}\right)\frac{s_{ij}^0}{\sigma_y} - \alpha\frac{r_{ij}}{2Gf(\xi_p)}\right]d\xi_p = de_{ij} - \frac{ds_{ij}^0}{2G}.$$

Multiply both sides of this equation with value s_{ij}^0 and pay attention to equation (9), we get

$$a\sigma_y d\xi_p = s_{ij}^0 de_{ij}, \tag{13}$$

in which

$$a = 1 + \frac{1}{3G}\frac{dr_{ij}}{d\xi_p}\frac{de_{ij}^p}{d\xi_p}. \tag{14}$$

Note that for materials used in practice, the inequality $\frac{dr_{ij}}{d\xi_p}\frac{de^p_{ij}}{d\xi_p} > 0$ is satisfied over the entire range of stress variation.

Consider that the proposed theory satisfies the following equality

$$de_{ij} = \frac{ds_{ij}}{2G} + \frac{3s^0_{ij}}{2\sigma_y}d\xi_p, \tag{15}$$

It follows from formula (13) that

$$d\xi_p = \frac{s^0_{ij}d\sigma_{ij}}{2G\sigma_y(a-1)}. \tag{16}$$

Finally, the deviator strain increment can be obtained from the equations (14)-(16)

$$de_{ij} = \frac{ds_{ij}}{2G} + \frac{s^0_{ij}s^0_{kl}d\sigma_{kl}}{L^2}, \tag{17}$$

where:

$$L^2 = \frac{4}{9}\sigma_y^2\frac{dr_{ij}}{d\xi_p}\frac{de^p_{ij}}{d\xi_p}.$$

The equality $s^0_{kl}ds_{kl}=s^0_{kl}d\sigma_{kl}$ is taken account of in formula (17).

Introduce auxiliary stress and strain vectors

$$\{\sigma\} = (\sigma_{11},\sigma_{22},\sigma_{33},\sigma_{12},\sigma_{13},\sigma_{23})^{\mathrm{T}};\ \{\varepsilon\} = (\varepsilon_{11},\varepsilon_{22},\varepsilon_{33},\gamma_{12},\gamma_{13},\gamma_{23})^{\mathrm{T}},$$

as well as active stress and back stress vectors

$$\{s^0\} = (s^0_{11},s^0_{22},s^0_{33},2s^0_{12},2s^0_{13},2s^0_{23})^{\mathrm{T}};\ \{r\} = (r_{11},r_{22},r_{33},r_{12},r_{13},r_{23})^{\mathrm{T}}.$$

Hence, in view of equality $d\varepsilon_{kk} = \frac{d\sigma_{kk}}{3K}$ formula (17) can be transformed to

$$d\{\varepsilon\} = [S_{ep}]d\{\sigma\}. \tag{18}$$

Here $[S_{ep}]$ – elastic-plastic compliance matrix, which is calculated by formula

$$[S_{ep}] = [S] + \frac{\{s^0\}\{s^0\}^T}{L^2}, \tag{19}$$

where $[S]$ – known elastic compliance matrix of isotropic solid

$$L^2 = \frac{4}{9}\sigma_y^2\frac{d\{r\}^T}{d\xi_p}\frac{d\{\varepsilon^P\}}{d\xi_p}.$$

The constitutive relations in formula (18) can be inverted and written in form

$$d\{\sigma\} = [D_{ep}]d\{\varepsilon\}. \tag{20}$$

Matrix $[D_{ep}]$- elastic-plastic stiffness matrix of isotropic solids. It can be found with the aid of known formula

$$([A] + [B])^{-1} = [A]^{-1} - \beta[A]^{-1}[B][A]^{-1}, \tag{21}$$

where $[A]$ – nonsingular matrix of the size $n \times n$; $[B] = \{u\}\{v\}^T$- matrix of the rank 1; $\{u\}$, $\{v\}$- vectors of size n; $\beta = (1+\alpha)^{-1}$, $\alpha = \{v\}^T[A]^{-1}\{u\}$.

Consider a particular case of the presented dependencies, which corresponds the loading process of composite pressure vessel subjected to an internal pressure. In this case, a plane stress state takes place in liner, in which $\sigma_{33} = \sigma_{13} = \sigma_{23} = 0$. Then from (18) we get

$$\left.\begin{aligned} d\varepsilon_{11} &= \left[\frac{1}{E} + \frac{(s^0_{11})^2}{L^2}\right]d\sigma_{11} + \left(-\frac{\nu}{E} + \frac{s^0_{11}s^0_{22}}{L^2}\right)d\sigma_{22} + 2\frac{s^0_{11}s^0_{12}}{L^2}d\sigma_{12},\\ d\varepsilon_{22} &= \left(-\frac{\nu}{E} + \frac{s^0_{11}s^0_{22}}{L^2}\right)d\sigma_{11} + \left[\frac{1}{E} + \frac{(s^0_{22})^2}{L^2}\right]d\sigma_{22} + 2\frac{s^0_{22}s^0_{12}}{L^2}d\sigma_{12},\\ d\gamma_{12} &= 2\frac{s^0_{11}s^0_{12}}{L^2}d\sigma_{11} + 2\frac{s^0_{22}s^0_{12}}{L^2}d\sigma_{22} + \left[\frac{1}{G} + \frac{4(s^0_{12})^2}{L^2}\right]d\sigma_{12}. \end{aligned}\right\} \tag{22}$$

Besides, during the loading process the following deformation also occurs

$$d\varepsilon_{33} = -\left[\frac{\nu}{E} + \frac{s^0_{11}(s^0_{11}+s^0_{22})}{L^2}\right]d\sigma_{11} - \left[\frac{\nu}{E} + \frac{s^0_{22}(s^0_{11}+s^0_{22})}{L^2}\right]d\sigma_{22} - 2\frac{s^0_{12}(s^0_{11}+s^0_{22})}{L^2}d\sigma_{12},$$

which provides reference for practical calculations. Here the expression for L^2 takes the following form

$$L^2 = \frac{4}{9}\sigma_y^2\left\{R - \frac{\alpha}{\sigma_T f(\xi_p)}\left[\left(\sigma_{11} - \frac{\sigma_{22}}{2} - \frac{3r_{11}}{2}\right)(2r_{11} + r_{22}) + \left(\sigma_{22} - \frac{\sigma_{11}}{2} - \frac{3r_{22}}{2}\right)(2r_{22} + r_{11}) + \right.\right.$$
$$\left.\left. +2(\sigma_{12} - r_{12})r_{12}\right]\right\}.$$

To obtain the back stresses r_{11}, r_{22}, r_{12}, it is advisable to apply the relation (11) at $i=j=1$; $i=j=2$; $i=1$, $j=2$. Increment of Odqvist parameter can be specified by formula

$$d\xi_p = \frac{2}{\sqrt{3}}\sqrt{\left(d\varepsilon_{11}^p\right)^2 + d\varepsilon_{11}^p d\varepsilon_{22}^p + \left(d\varepsilon_{22}^p\right)^2 + \frac{\left(d\gamma_{12}^p\right)^2}{4}}.$$

Expression for loading surface (10) takes following form

$$(s_{11} - r_{11})^2 + (s_{22} - r_{22})^2 + (s_{11} - r_{11})(s_{22} - r_{22}) + (s_{12} - r_{12})^2 = \frac{\sigma_y^2}{3}.$$

When calculating a multilayer composite material containing an isotropic layer, equations (22) should be reversed. This operation can be done with the aid of formula (21). In this case introduce auxiliary values $x_1 = \frac{s_{11}^0}{L}, x_2 = \frac{s_{22}^0}{L}, x_3 = 2\frac{s_{12}^0}{L}$. Then the equality (22) can be rewritten as follows

$$\left.\begin{aligned} d\varepsilon_{11} &= \left(\frac{1}{E} + x_1^2\right)d\sigma_{11} + \left(-\frac{\nu}{E} + x_1x_2\right)d\sigma_{22} + x_1x_3 d\sigma_{12}, \\ d\varepsilon_{22} &= \left(-\frac{\nu}{E} + x_1x_2\right)d\sigma_{11} + \left(\frac{1}{E} + x_2^2\right)d\sigma_{22} + x_2x_3 d\sigma_{12}, \\ d\gamma_{12} &= x_1x_3 d\sigma_{11} + x_2x_3 d\sigma_{12} + \left(\frac{1}{G} + x_3^2\right)d\sigma_{12} \end{aligned}\right\}.$$

The inverse equalities have the following forms

$$\left.\begin{aligned} d\sigma_{11} &= \left(\frac{E}{1-\nu^2} - y_1^2\right)d\varepsilon_{11} + \left(\frac{E\nu}{1-\nu^2} - y_1y_2\right)d\varepsilon_{22} + y_1y_3 d\gamma_{12}, \\ d\sigma_{22} &= \left(\frac{E\nu}{1-\nu^2} - y_1y_2\right)d\varepsilon_{11} + \left(\frac{E}{1-\nu^2} - y_2^2\right)d\varepsilon_{22} + y_2y_3 d\gamma_{12}, \\ d\sigma_{12} &= y_1y_3 d\varepsilon_{11} + y_2y_3 d\varepsilon_{22} + (G + y_3^2)d\gamma_{12}, \end{aligned}\right\}. \qquad (23)$$

Here the following notations are introduced

$$y_1 = \frac{E(s_{11}^0 + \nu s_{22}^0)}{C(1-\nu^2)}, \quad y_2 = \frac{E(s_{22}^0 + \nu s_{11}^0)}{C(1-\nu^2)}, \quad y_3 = \frac{2G}{C}s_{12}^0$$

$$C = \sqrt{L^2 + \frac{E}{1-\nu^2}[(s_{11}^0)^2 + (s_{22}^0)^2 + 2\nu s_{11}^0 s_{22}^0] + 4G(s_{12}^0)^2}.$$

In conjunction with equation (23) elastic-plastic stiffness matrix of liner material will take the following form

$$\left[D_{\text{м}}^{ep}\right] = \begin{bmatrix} \frac{E}{1-\nu^2} - y_1^2 & \frac{E\nu}{1-\nu^2} - y_1y_2 & -y_1y_3 \\ \frac{E\nu}{1-\nu^2} - y_1y_2 & \frac{E}{1-\nu^2} - y_1^2 & -y_2y_3 \\ -y_1y_3 & -y_2y_3 & \frac{E}{2(1+\nu)} - y_3^2 \end{bmatrix}.$$

A calculation algorithm is developed for modelling a cylindrical part subjected to an internal pressure, which is based on the proposed constitutive equations. In accordance with it the elastic-plastic stiffness matrix of multilayer material can be found by formula

$$\left[G_c^{ep}\right] = \left[D_{\text{м}}^{ep}\right]\delta_{\text{м}} + [D_{\text{км}}]\delta_{\text{км}}. \qquad (24)$$

Here matrix $[D_{\text{км}}]$ is determined from equality (4). If matrix failure takes place in composite layer, then the monotropic model should be used for unidirectional composite material, which yields

$$[D_{\text{км}}] = E_1\begin{bmatrix} \delta_c cos^4\varphi & \delta_c sin^2\varphi cos^2\varphi & 0 \\ \delta_c sin^2\varphi cos^2\varphi & \delta_c sin^4\varphi + \delta_k & 0 \\ 0 & 0 & \delta_c sin^2\varphi cos^2\varphi \end{bmatrix},$$

where E_1 – longitudinal modulus of the unidirectional composite material, $\delta_c = h_c/h_{\text{км}}$, $\delta_{\text{к}} = h_{\text{к}}/h_{\text{км}}$. When analyzing the strength properties of cylindrical part for the unidirectional composite material, the Hanshin-Rotem criterion [8] is applied, it follows

$$\sigma_{11k\text{м}} = F_{+1}. \qquad \frac{\sigma_{22k\text{м}}{}^2}{F_{+2}} + \frac{\sigma_{12k\text{м}}{}^2}{F_{12}} = 1,$$

where F_{+1} - longitudinal tensile ultimate stress, F_{+2} - transverse tensile ultimate stress, F_{12} - shear ultimate stress.

The calculation algorithm is based on the step loading methods and variable parameters of elasticity. The computation is achieved in the *MatLab* package.

3. Numerical example

Examples in conjunction with above presented mathematical model are given which analyze the multilayer composite material, consider the cylindrical part of a pressure vessel type 3, which is discussed in works [1,2]. External radius of cylindrical part $R_m = 182.2\ mm$. Composite carbon-epoxy layer has the following mechanical properties: $E_1 = 130\ GPa$; $E_2 = 11\ GPa$; $\nu_{12} = 0.27$; $G_{12} = 4.7\ GPa$; $F_{+1} = 1600\ MPa$; $F_{+2} = 23.8\ MPa$; $F_{12} = 33.3\ MPa$. Angle of the helical layer $\varphi_c = 13°$; thickness of the helical layer $h_c = 0.6\ mm$; thickness of the circumferential layer $h_K = 1.2\ mm$.

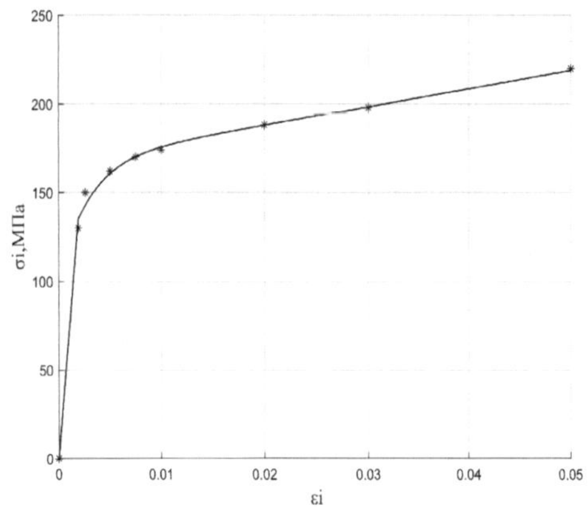

Figure 2. Stress-Strain diagram of aluminum: ▬▬ – approximation, which is based on the proposed theory of plasity with translational hardening; * - experimental data.

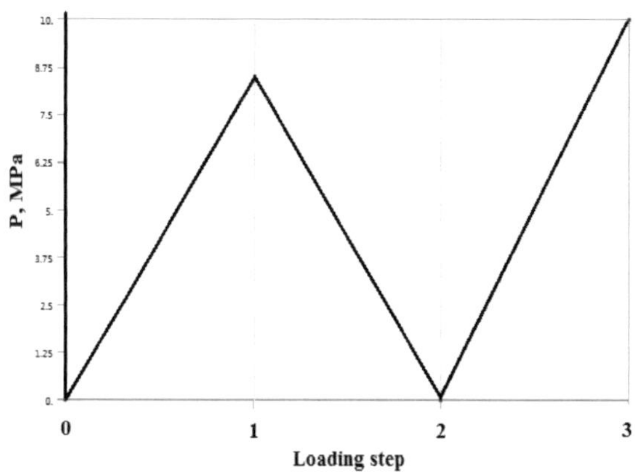

Figure 3. Loading process of composite pressure vessel.

The liner of studied pressure vessel consists of aluminum alloy *AMg-6* with thickness $h_m = 2.2\ mm$. The alloy has the following characteristics: Young's modulus $E = 70.6\ GPa$; Poisson's ratio ν =0.3; yield stress $\sigma_y = 135\ MPa$. The material parameters of alloy are as follows: $\alpha = 400$; $\beta = 30$; $R = 15000\ MPa$. The values are chosen from the condition of the best agreement between theoretical and experimental data.

The stress-strain diagram of the aluminum alloy *AMg-6* is shown in Fig. 2. It can be approximated with the equality, which is obtained from equation (12). Under uniaxial tension the diagram can be taken in the following form

$$\sigma_{11} = \sigma_y + \frac{R}{\alpha+\beta}\left[\left(1+\beta\varepsilon_{11p}\right) - \left(1+\beta\varepsilon_{11p}\right)^{\frac{-\alpha}{\beta}}\right];$$

$$\varepsilon_{11} = \frac{\sigma_{11}}{E} + \varepsilon_{11p}.$$

The finite element method is also employed for comparison. In previous paper, a model of cylindrical part of composite pressure vessel is developed in *ANSYS-WORKBENCH* software package, which is simulated by shell finite elements. In this case, the Prandtl-Reuss relations are used to describe the plastic properties of liner material.

The pressure P_y is 3.41 MPa beyond which the liner material experience plastic deformation. It can be found with the aid of formula (8). Then the vessel is loaded up to the pressure $P_a = 8.5\ MPa$, unloaded to zero pressure and reloaded to the pressure $P_{\text{Max}} = 10\, MPa$ according to which the deformation of the cylindrical part is investigated. The loading process is shown in Fig 3.

Dependencies of axial and circumferential strains in cylindrical part of the vessel are presented in Fig 4, 5, respectively. The presented results are derived on basis of relations of plastic models with isotropic and translational hardening and show that the matrix failure first occurs in helical layer at $P_p^c = 3.84\, MPa$. This pressure corresponds to square in Fig 4, 5. And at $P_p^{\kappa} = 4.76\, MPa$ (shown in Fig 4, 5 with triangle) the matrix failure takes place in the circumferential layer. Comparing the obtained results with experimental data, we find that the Prandtl-Reuss model with isotropic hardening is not able to describe the Bauschinger effect during unloading process and the obtained corresponding residual strains exceed the experimental data.

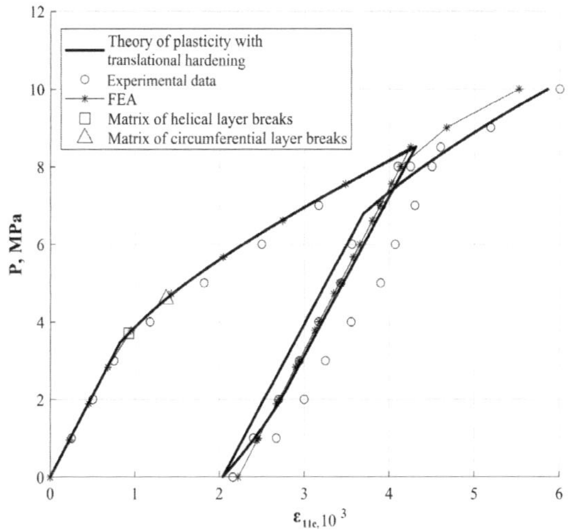

Figure 4. Dependencies of the axial strains in cylindrical part of the vessel on internal pressure.

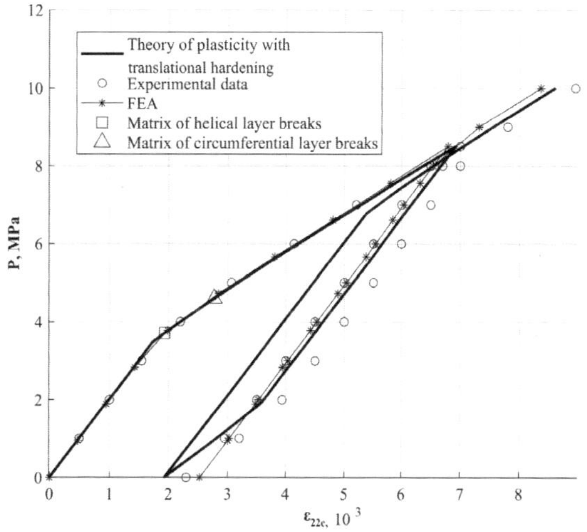

Figure 5. Dependencies of the circumferential strains in cylindrical part of the vessel on internal pressure.

As can be seen in Fig 6, after unloading compressive residual stresses remain in liner material which can cause the thin-walled liner to buckling. The obtained results show that it is necessary to take the translational hardening of liner material into account, since the difference between the residual stresses found according to the two presented theories can reach $\approx 30\%$.

Noted that the alternating stresses in liner material have essential aspects for analyzing low-cycle fatigue of a vessel. The amplitude range can be significant. Under cyclic loading of a composite pressure vessel with load-sharing liner, one should expect that the smaller the amplitude of stress variation, the higher the number of loading cycles till failure happens. In accordance with the forging fact the accurate prediction of stress variation in liner material supposed to be relevant.

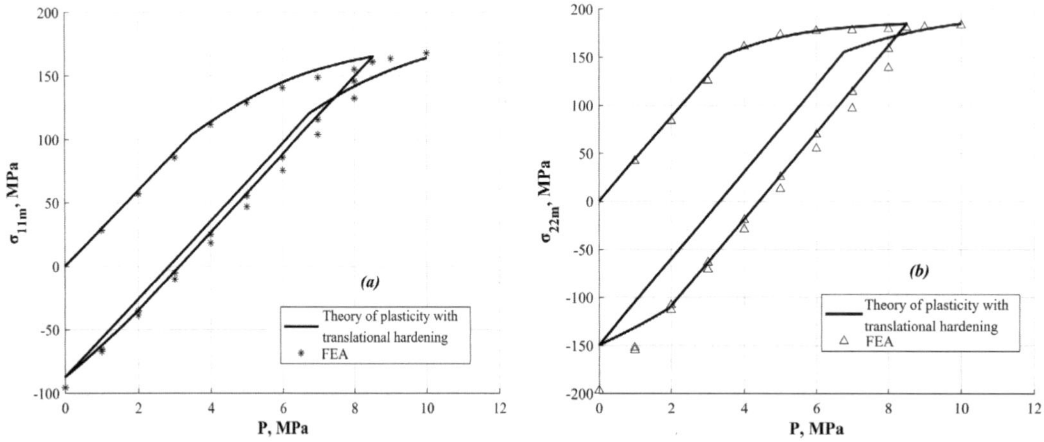

Figure 6. Dependencies axial ***(a)*** and circumferential ***(b)*** stresses in liner material on internal pressure.

The calculation results graphically presented in Fig 7 show that after unloading the longitudinal stresses acting in circumferential and helical composite layers are tensile. After unloading the obtained stresses in liner material which correspond the plastic model with translational hardening, are lower than the stresses which obtained on the basis of plastic model with isotropic hardening. Therefore, using the above proposed relations for metallic liner, we can clarify the stress state in monolayers of a multilayer composite material. This effect plays an important role in both static and low-cycle strength evaluations of the structure.

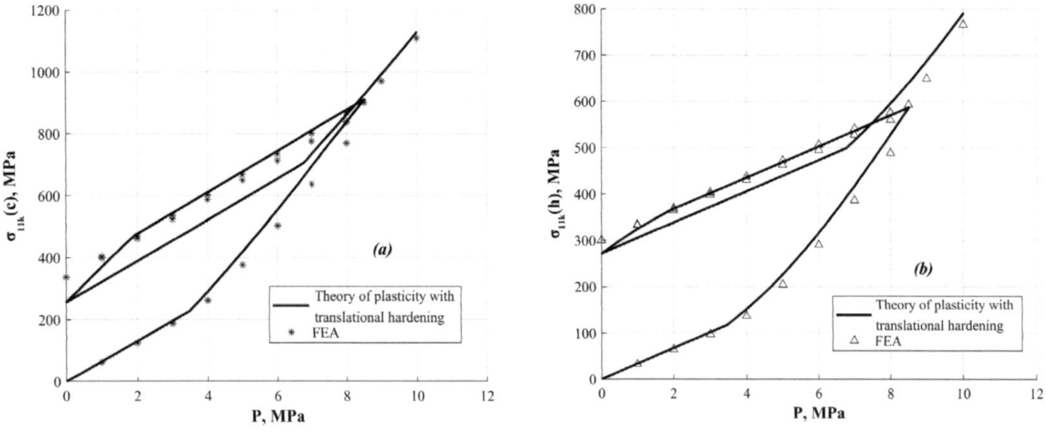

Figure 7. Dependencies longitudinal stress σ_{11k} in the circumferential ***(a)*** and helical ***(b)*** layers on internal pressure

4. Conclusions

In present work a solution for analyzing the cylindrical part of a metallic-composite pressure vessel with a load-sharing liner is presented, in which the constitutive relations based on the theory of plasticity of isotopic solids with translational hardening are used to describe the plastic properties of liner material. A fibrous monolayer, which is part of a multilayer composite material, considered to be an orthotropic linear elastic solid. To determine the ultimate strength of a monolayer, the Hashin-Rotem criterion is considered. The proposed theoretical relationships are written with the help of the elastic-plastic stiffness matrix for a multilayer material. The developed constitutive relations are shown to not only qualitatively describe the formation of residual stresses and strains during unloading process, but also predict the properties of multilayer material under cyclic loading. The proposed constitutive relations appear to be

IOP Conf. Series: Materials Science and Engineering 1191 (2021) 012020 doi:10.1088/1757-899X/1191/1/012020

convenient for solving practical works. The calculation for cylindrical part of metallic-composite pressure vessel has been validated by the comparison of the results governed by relations of classical Prandtl-Reuss flow rule and proposed theory of plasticity with translational hardening. Meanwhile, the residual stresses and strains after unloading and deformation during subsequent loading are analyzed. It is shown that the application of the solution based on the theory of plasticity with translational hardening leads to better agreement between theoretical and experimental data, particularly during the unloading reloading processes. The proposed mathematical model allows one to adequately predict the bearing capacity of metallic-composite pressure vessel under low-cycle loading.

References

[1] Vasiliev V.V. Composite pressure vessels. Analysis, design, manufacturing. 2009 *Blacksburg. Bull Ridge Publishing*. 704p
[2] Vasiliev V.V., Moroz N.G. Composite pressure vessels. Analysis, design, manufacturing. 2015 *M.: Mashinostroinie* 373p (*in Russian*)
[3] Sarbayev B.S. 1994. *Mechanical of Solids* **1** 65-72
[4] GOST 33986-2016. Technical requirments and methods of test. 2017. *M.: Standartinform* 15p (*in Russian*)
[5] Trifonov O. V., Cherniy V.P. 2014. *Journal of Pressure Vessel Technology* **136 051202** 1-8
[6] Burov A.E., Burova O.G. 2019. *Journal of Physics: Conference Series* **1441** 1-7
[7] Burov A.E., Lepihin A.M., Makhutov N.A. 2017. *Strength of Materials* **49** 666-675
[8] Hashin Z. 1980 *J. Appl. Mech* **47** 329–334.

Computer simulation of unloading a dump truck on an inclined platform.

Yu N Baryshnikov

Department of Theoretical Mechanics, Bauman Moscow State Technical University, Moscow, 105005, Russia

E-mail: mhts@list.ru

Abstract. The article deals with the problem of loss of stability of a dump truck when unloading on an inclined platform. The main task of the study is to determine the maximum permissible angle of inclination of the site. The analysis of modern methods of solving the problem is given. A mathematical model of a dump truck when unloading on a site with a longitudinal slope is developed. In the case of cargo sticking to the platform, an analytical formula is obtained for calculating the critical angle of inclination of the platform. When the cargo is moving, the procedure of step-by-step change of parameters is proposed. Based on the obtained ratios, a program for calculating the longitudinal stability of the dump truck is compiled. The results of assessing the longitudinal stability of two KAMAZ dump trucks are presented.

1. Introduction

Overturning of the car (loss of stability) can occur both in the longitudinal and transverse plane. In the longitudinal plane: when braking sharply on a steep descent, when driving on a steep ascent, when reversing into a ravine or river. In the transverse plane: when driving at high speed on sharp turns, when the side skid abruptly stops as a result of hitting an obstacle with the rear wheel, when the load is unevenly loaded or moved on a turn, etc. [1].

These cases are also found in the operation of dump trucks [2]. In addition, for this class of vehicles, there is a risk of overturning the vehicle when unloading. Overturning occurs during unloading on inclined platforms or when the rear wheels sink into the ground. In this regard, the main task in assessing the stability of the dump truck is to determine the maximum permissible (critical) angle of inclination of the unloading site.

2. Modern methods of assessing sustainability.

Traditionally, the critical angle of the site is determined when testing full-scale samples of cars on special stands. Our country has developed state standards that regulate such tests [3]. Similar tests are carried out abroad. For example, experimental studies of the stability of heavy-duty vehicles with a lateral (transverse) slope are performed at the company Fliegl Fahrzeugbau [4]. At the Schmitz Cargobull company, all dump semi-trailers are tested for rollover resistance both backward and sideways at the development stage with the help of a unique test bench [5]. However, this approach requires significant costs for conducting expensive tests and increases the time of finishing cars.

Another approach to solving this problem is based on the use of computational research methods. In particular, at the Chelyabinsk Polytechnic Institute, the critical angle of the unloading platform was calculated on a complete finite element model of the KAMAZ-6520 car [6]. The model took into account the suspension design, the non-linearity of the elastic suspension elements and tires, as well as the possibility of plastic deformation of the bearing system during torsion. However, the development and debugging of such a model requires not only considerable time, but also the well-coordinated work of a whole team of highly qualified engineers. In addition, the results obtained require experimental verification and are the subject of serious scientific research.

At the same time, there is a need to develop a relatively simple, engineering method for calculating the critical slope angle of the site to assess the stability of the dump truck.

3. Longitudinal stability of the dump truck.

Consider the scheme of unloading a dump truck on a longitudinal slope (Fig.1).

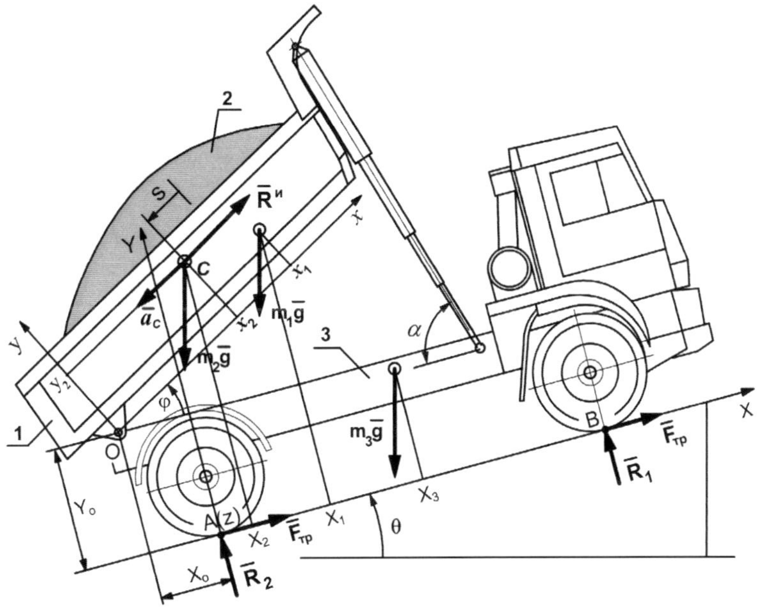

Figure 1. Calculation scheme of dump truck unloading.

Here m_1 and m_2 – the mass of the platform 1 and the load 2; m_3 – the mass of the chassis 3; φ – the angle of lifting of the platform; θ – the angle of inclination of the unloading platform; s – movement of the load during unloading; R_1 and R_2 –total reactions on the wheels of the front and rear axles of the car; $AB = L$ – wheelbase.

Let's use the Dalembert principle [7]:

$$\sum M_Z(m_\kappa \bar{g}) + \sum M_Z(\bar{R}_\kappa) + \sum M_Z(\bar{\Phi}_\kappa) = 0 , \tag{1}$$

Let the load be a monolithic block, symmetrical with respect to the vertical plane of symmetry Oxy of the car. We bring all the inertia forces of the load $(\bar{\Phi}_\kappa)$ to its center of mass C in the form of the main vector $\bar{R}^{ин}$ and the main moment $\bar{L}_C^{ин}$. Then, as shown in [8], the main moment of the inertia forces of the load $\bar{L}_C^{ин} = -J_{Cz}\bar{\varepsilon}$ is zero. In turn, the main vector of inertia forces $\bar{R}^{ин}$ is directed in the opposite direction of the movement of the load (Fig. 1), and its modulus is equal to: $R^{ин} = m_2 \ddot{s}$.

Taking into account the obtained relations, equation (1) is written as:

$$-\sum_{\kappa=1}^{3} m_\kappa g \cos\theta \cdot X_\kappa + \sum_{\kappa=1}^{3} m_\kappa g \sin\theta \cdot Y_\kappa + R_1 \cdot L - R^{ин}(y_2 + Y_0 \cos\varphi + X_0 \sin\varphi) = 0, \tag{2}$$

Here, the coordinates of the centers of mass of the platform 1 and the load 2 in the stationary $(X_\kappa; Y_\kappa)$ and mobile $(x_\kappa; y_\kappa)$ coordinate systems are related to each other by the following relations:

$$X_\kappa = x_\kappa \cos\varphi - y_\kappa \sin\varphi - X_O; \tag{3}$$
$$Y_\kappa = x_\kappa \sin\varphi + y_\kappa \cos\varphi + Y_O, \qquad (\kappa = 1, 2) \tag{4}$$

If the angle of inclination of the platform is equal to the critical angle θ^*, the front wheels of the dump truck break away from the support. In this case, the total reaction R_1 on the front wheels will become zero and equation (2) will take the form:

$$-g\cos\theta^* \sum_{\kappa=1}^{3} m_\kappa X_\kappa + g\sin\theta^* \sum_{\kappa=1}^{3} m_\kappa Y_\kappa - m_2\ddot{s}(y_2 + Y_0\cos\varphi + X_0\sin\varphi) = 0. \tag{5}$$

In the case of sticking (freezing) of the load to the platform $(s = 0)$ from equation (5), we find the critical angle of inclination of the platform:

$$\theta^* = arctg\left(\frac{\sum_{\kappa=1}^{3} m_\kappa X_\kappa}{\sum_{\kappa=1}^{3} m_\kappa Y_\kappa}\right) \tag{6}$$

After substituting the relations (3) and (4) in the formula (6), we get:

$$\theta^* = arctg\left(\frac{(m_1x_1 + m_2x_2)\cos\varphi - (m_1y_1 + m_2y_2)\sin\varphi - (m_1 + m_2)X_0 + m_3X_3}{(m_1x_1 + m_2x_2)\sin\varphi + (m_1y_1 + m_2y_2)\cos\varphi + (m_1 + m_2)Y_0 + m_3Y_3}\right) \tag{7}$$

Thus, according to the formula (7), already at the design stage, it is possible to calculate the critical slope angle of the platform and make an assessment of the stability of the dump truck during unloading.

In the case of cargo movement $(s > 0)$, the following algorithm for solving the problem is developed to calculate the critical angle of inclination. We set the initial values of the angles $\theta = 0^\circ$ and $\varphi = 0^\circ$, the coordinates $(x_\kappa; y_\kappa, \ \kappa = 1, 2)$ of the centers of mass of the platform and the load in the initial position, and other parameters. Then discretely, we increase the angle φ of lifting of the platform. When the total angle $(\varphi + \theta)$ becomes equal to the angle of the natural slope $(\approx 30^\circ)$, we simulate the movement of the load. To do this, at each step of changing the angle φ, we calculate all the parameters included in equation (5). In particular, the acceleration $a_C = \ddot{s}$ of the center of mass of the load is found from the differential equation of its motion:

$$m_2\ddot{s} = m_2 g(\sin(\varphi + \theta) - f\cos(\varphi + \theta)), \tag{8}$$

where f – is the coefficient of friction of sliding the load on the platform.

Integrating equation (8), we find the displacement s of the load. From the relations (3) and (4), we calculate the new coordinates of the center of mass of the platform and the load. All the found parameters are substituted into equation (5). If the sum of the moments in equation (5) is greater than or equal to zero, the car will overturn. This completes the calculation and fixes the value of the critical angle of inclination of the platform. Otherwise, we increase the angle by one degree and repeat the unloading process. Thus, by consistently changing the angle of inclination of the platform, we determine its critical value in the case of cargo movement.

4. Calculation of the stability of KAMAZ dump trucks.

To assess the longitudinal stability based on the formula (7), two models of KAMAZ dump trucks were selected. Their technical characteristics are presented in Tables 1 and 2.

Table 1. Technical characteristics of KAMAZ dump trucks.

Complete set of the dump truck	Front axle load, kg		Rear axle load, kg		Weight, kg	
Model	1	2	1	2	1	2
Complete	5300	7500	9500	13000	14800	20500
Curb weight	3450	4525	3450	4200	6900	8725
Chassis	3400	4025	2050	2800	5450	6825

Note: Model: 1 - KAMAZ-43255; 2 - KAMAZ-53605.

Table 2. Parameters of KAMAZ dump trucks.

The parameters of dump truck	KAMAZ-43255	KAMAZ-53605
Cargo weight, kg	7750	11700
Wheelbase, mm	3500	3950
Coordinates of the pivot support, mm	$X_o = 950$ $Y_o = 1050$	$X_o = 960$ $Y_o = 1065$
Maximum lifting angle of the platform	55°	50°
Lifting time, sec.	30	30

Based on the parameters of the cars (Table 1 and 2) the initial data for calculating the critical slope angle of the platform are obtained. The results of calculations of critical angles when changing the lifting angle of a loaded platform are shown in Figure 2.

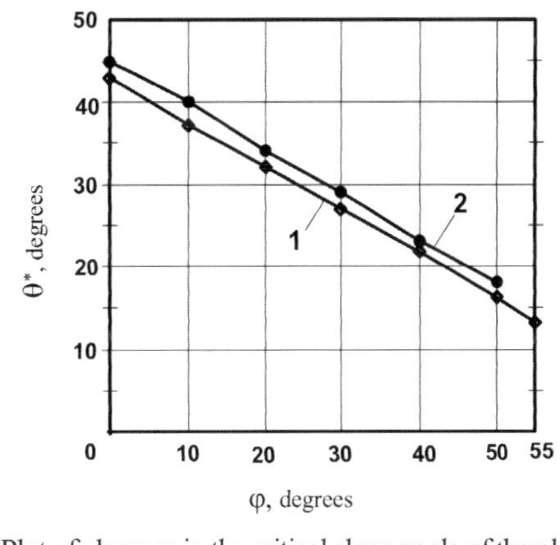

Figure 2. Plot of changes in the critical slope angle of the platform.
1 - KAMAZ-43255; 2 - KAMAZ-53605

From the graphs shown, it can be seen that these models of KAMAZ dump trucks have high longitudinal stability. So when lifting a loaded platform to the maximum angle $(\varphi = 50^\circ)$, the KAMAZ-53605 maintains stability on the platform with an angle of inclination $\theta = 17^\circ$, and the KAMAZ-43255-on the platform with an angle $\theta = 13^\circ$.

In the case of a moving load, the critical angle of inclination of the platform will be greater than the obtained values. This is due to the forces of inertia of the load, which tend to keep the dump truck in a stable position.

Note that the design scheme of the dump truck did not take into account the elastic properties of the suspension and tires. However, this can be done within the framework of this model. Indeed, when lifting the platform, the load on the rear axle increases. As a result of the additional deflection of the rear suspension springs and tires, the dump truck has its own angle of longitudinal inclination. Calculations have shown that the maximum angle of proper inclination (at the end of the platform lift) is 2-4 degrees. Therefore, the actual critical slope angle of the platform will be less than the calculated value by the specified amount. A similar angle of inclination of KAMAZ dump trucks occurs when the rear wheels sink into the ground by 800 mm and 500 mm, respectively.

5. Conclusions

The proposed method is an effective tool for express analysis of dump truck stability. Using formula (7) to calculate the critical angle of inclination of the site for a dump truck with a stuck load, it is possible to carry out multivariate calculations at an early stage of vehicle design in order to select the optimal design solution. With this approach, the timing of the development of cars and the cost of conducting expensive tests are significantly reduced.

References

[1] Kravets V. N. Theory of the automobile. N. Novgorod: NSTU named after R. E. Alekseev, 2013. 413 p
[2] Belokurov V. N., Gladkov O. V., Zakharov A. A., Melik-Sarkisyants A. S. Autos-dump trucks / ed. by A. S. Melik-Sarkisyants. Moscow: Mashinostroenie, 1987. 216 p
[3] GOST 31507-2012 Motor vehicles. Controllability and stability. Technical requirements. Test methods. Moscow: Standartinform, 2013. 51p
[4] Drong V. I., Dubinin V. V., Ilyin M. M. et al. Course of theoretical mechanics: textbook for universities//Drong V. I., Dubinin V. V., Ilyin M. M. et al. Edited by K. S. Kolesnikov, V. V. Dubinin. 5th ed., ispr. - M.: Publishing House of the Bauman Moscow State Technical University, 2017. 584 p
[5] Test for the stability of dump bodies and chassis / / Access mode: http://www.fliegl.ru/news/13.htm
[6] Karasev A. Ease from Schmitz Cargobul. 2010. No. 5. p. 48, 49 Access mode: http://www.autotruck-press.ru/archive/number93/article8
[7] The calculation of the lateral resistance while unloading the truck of "KAMAZ-6520". 2009 // Laboratory Of Vibration
[8] Baryshnikov Yu. N. Stability of a dump truck during unloading / / Machines and installations: design, development and operation, Bauman Moscow State Technical University, Electron. journal 2015, no. 3, pp. 51-61, DIO: 10.7463/aplts. 0315. 0783168

IOP Conf. Series: Materials Science and Engineering 1191 (2021) 012022 doi:10.1088/1757-899X/1191/1/012022

Promising methods for predicting the residual life of polymer composites in refined variational statements under extreme environmental influences

Gusev E.L.[1,2*], Bakulin V.N.[3], Chernykh V.D.[1]

[1]Institute of oil and gas problems of the Federal research center «Yakut scientific center» Siberian Branch RAS (IOGP FRC «YSC» SB RAS, Petrovsky Str. 2 , Yakutsk, 677980, Russia
[2]Institute of mathematics and computer science of the North- Eastern Federal University (IMCS NEFU) , Kulakovsky str. 48, Yakutsk, 677013, Russia
[3]Institute of applied mechanics RAS (IAM RAS), Leningradsky Ave 7, Moscow, 125040, Russia

* E-mail: elgusev@mail.ru

Abstract. Modifications of the methods for finding the global extremum of multi-parametric performance indicators are developed, taking into account the structural features of the quality assessment criteria used in predicting the defining characteristics of composites. The article analyzes promising ways to increase the potential opportunities associated with the improvement of methods for constructing optimal directions for finding the absolute minimum of efficiency indicators associated with solving problems of predicting the residual life, reliability, and durability of polymer composites under the influence of extreme environmental factors and operational loads . The comparative analysis of existing approaches and their possible modifications allowed us to conclude that the most promising opportunities for the effective construction of globally optimal solutions in refined variational formulations of inverse forecasting problems are associated with the development of a methodology for effectively narrowing the search areas containing globally optimal solutions. The results of comparative computational experiments are presented.

1. Introduction

In recent decades, polymer composite materials, composite structures, and composite coatings have become increasingly common in modern areas of technology that determine scientific and technological progress, such as aviation and space technology, shipbuilding, oil and gas industry systems, and modern equipment operating in extreme conditions of the Arctic and Subarctic zones [1-4]. Due to the growing use of polymer composites in engineering, the problem of predicting the reliability, durability, and residual life of composite materials and structures made of them is becoming very relevant [5-8]. Studies related to the possibility of developing effective methods for constructing the most promising directions for searching for globally optimal solutions in variational formulations of inverse problems of predicting the residual resource of composites, taking into account the established qualitative regularities of the structure of optimal solutions, are carried out.

2. Statement of the problem

As a defining characteristic of a composite structure, its residual life can be considered [9-11]. Models describing the dependence of changes in one of the defining characteristics (for example, the strength R of polymer composites) under the influence of extreme climatic factors can generally be presented in the form of functional dependencies of the following type [12-14]:

$$R\left(u_1,u_2,...,u_n;t\right)=R_0+\Phi\left(u_1,u_2,...,u_n;t\right).$$

$$\Delta R\left(u;t\right)=\Phi\left(u;t\right),\quad \Delta R=R-R_0. \tag{1}$$

Let us denote $\Delta\overline{R}_t^э$ - the experimental mean values of the function $\Delta R=R-R_0$ on experimental samples after the t-th year (t=1,2,..., m). The parameters of the optimal forecasting model $u^*=(u_1^*,u_2^*,...,u_m^*)$ give an extreme value to the special efficiency criterion J, which is a complex multiparametric function of the parameters of the forecasting problem:

$$J(u)=J\left(\left\{\Delta R\left(u,t\right)\right\}_{t=1}^{m};\ \left\{\Delta\overline{R}_t^э\right\}_{t=1}^{m}\right). \tag{2}$$

As a criterion of efficiency as a complex function of the parameters of the forecasting problem of the form (2), it is advisable to choose the standard deviation of the approximation of experimental data $R_t^э$ from the theoretical values of R(u, t):

$$J(u_1,u_2,...,u_n)=S^2=\frac{1}{m}\sum_{t=1}^{m}\left(R(u_1,u_2,...,u_n;t_i)-R_{t_i}^э\right)^2 . \tag{3}$$

In the case when the experimental data sufficiently adequately reflect the structure of the dependence of changes in the defining characteristics of composite materials on the impact of extreme environmental factors and operational loads, and the experimental data are obtained with minor errors that do not significantly distort the patterns of behavior of real dependencies, the problem of reconstructing the parameters of composite materials models from the influence of extreme environmental factors and operational loads can be reduced to finding the absolute minimum of the quality function (4) in the extreme problem:

$$J(u^*)=\min_u J(u). \tag{4}$$

The problem of reconstructing the parameters of models of changes in the residual life of polymer composites from the influence of extreme environmental factors and operational loads can be reduced to solving the extreme problem (4). The vector of parameters $u^*=\left(u_1^*,u_2^*,...,u_n^*\right)$ that deliver a minimum to the efficiency indicator J(u) (4) determines the optimal predicted dependence of changes in the defining characteristics of the polymer composite on the influence of extreme environmental factors and operational loads.

IOP Conf. Series: Materials Science and Engineering 1191 (2021) 012022 doi:10.1088/1757-899X/1191/1/012022

3. Materials and methods of research

The comparative analysis of existing approaches and their possible modifications allowed us to conclude that the most promising opportunities for the effective construction of globally optimal solutions in variational formulations of inverse forecasting problems are associated with the development of a methodology for effectively narrowing the search areas containing globally optimal solutions. At the same time, the developed methodology is based on methods related to the theoretical justification of the admissibility of schemes for narrowing the set of acceptable options, in which the desired globally optimal solution continues to be found on the resulting set of smaller dimensions [15-16].

Based on the obtained results and computational experiments, a method for effectively narrowing the set of acceptable variants of multiparametric models of various complexity M to a compact set M* containing an optimal model for predicting optimal complexity is developed. Based on a constructive study of the structure of performance indicators in the studied variational formulations of inverse problems of predicting the residual resource of composites, it is established that optimal solutions that deliver an absolute minimum to the corresponding performance indicators can additionally satisfy a certain system of relations.

The development of a methodology for identifying such systems of additional relations can significantly reduce the dimension of the initial variational statements of inverse problems of predicting the residual resource of composites and, accordingly, significantly narrow the initial set of acceptable options to a narrow compact set M*, and, accordingly, significantly narrow the initial set of acceptable options for forecasting models, on which to build a globally optimal solution corresponding to the absolute minimum of the efficiency indicator, effective methods of non-local optimization can be developed. Functional relations are established, which can additionally be satisfied by the desired globally optimal solutions in variational formulations of inverse problems of forecasting the residual resource. The established functional relations can generally be written as:

$$g_j\left(u_1, u_2, \ldots, u_n\right) = 0, \quad (j = 1, 2, \ldots, q; q < n). \tag{5}$$

The constructive analysis of the studied variational statements of inverse problems of predicting the residual resource of polymer composites allowed us to establish that the necessary conditions of the extremum for performance indicators of the form (3) for a number of parameters simultaneously coincide with sufficient conditions and these relations determine only globally optimal solutions, which are the desired solutions in the problems under consideration. We denote such parameters by

$u_{j_1}, u_{j_2}, \ldots, u_{j_q}$. The necessary minimum conditions for this selected system of parameters

$u_{j_1}, u_{j_2}, \ldots, u_{j_q}$ can be taken as a system of functional relations (5), which can serve as a basis for the subsequent significant reduction in the dimension of the original forecasting problem and simplification of its structure. According to this

$$g_i(u) = \frac{\partial J\left(u_1, u_2, \ldots, u_n\right)}{\partial u_{j_i}}, \ (i = 1, 2, \ldots, q). \tag{6}$$

We calculate on the basis of solving a system of algebraic equations, which lead to the necessary and sufficient conditions of the minimum parameters $u_{j_1}, u_{j_2}, \ldots, u_{j_q}$ through the remaining system of r=n-q parameters $u_l \ (l = 1, 2, \ldots, n;\ l \neq u_{j_i}, i = 1, 2, \ldots, q)$:

IOP Conf. Series: Materials Science and Engineering 1191 (2021) 012022 doi:10.1088/1757-899X/1191/1/012022

$$u_{j_i} = \varphi_i\left(u_1, u_2, \dots, u_{j_1-1}, u_{j_1+1}, \dots, u_{j_q-1}, u_{j_q+1, \dots}, u_{n-1}, u_n\right), \quad i = 1, 2, \dots, q. \tag{7}$$

As a result, we obtain a variational problem of much smaller dimension and correspondingly simpler structure of the form

$$\begin{gathered} \Phi\left(u_{l_1}, u_{l_2}, \dots, u_{l_r}\right) \Rightarrow \min, \\ \left(u_{l_1}, u_{l_2}, \dots, u_{l_r}\right) \in R \subset E^r,\ 1 \le u_{l_k} \le n, k = 1, 2, \dots, r, \\ \left(u_{l_k} \ne u_{j_i}, k = 1, 2, \dots, r ; i = 1, 2, \dots, q \right). \end{gathered} \tag{8}$$

In these notations

$$\begin{gathered} \Phi\left(u_{l_1}, u_{l_2}, \dots, u_{l_r}\right) = \\ = J\left(\varphi_1\left(u_{l_1}, u_{l_2}, \dots, u_{l_r}\right), \dots, \varphi_q\left(u_{l_1}, u_{l_2}, \dots, u_{l_r}\right), u_{l_1}, u_{l_2}, \dots, u_{l_r}\right). \end{gathered} \tag{9}$$

Effective methods of nonlocal optimization are developed for the formulated auxiliary variational problem (8), which has a significantly smaller dimension and simpler structure compared to the original variational problem, taking into account the established features of its structure. A statement is formulated that justifies the approach associated with an effective reduction in the dimension of the initial variational problem of predicting the residual resource of composites.

Statement. Let the functions $g_i\left(u_1, u_2, \dots, u_n\right), (i = 1, 2, \dots, q)$ be continuous in their arguments, there exist and are continuous partial derivatives $\partial g_i / \partial u_j$, $(i = 1, 2, \dots q; j = 1, 2, \dots, n)$; $u^* = \left(u_1^*, u_2^*, \dots, u_n^*\right)$ there is an optimal solution to the original prediction problem, and the condition is met

$$\begin{gathered} rang\left[\nabla g_1\left(u^*\right), \nabla g_2\left(u^*\right), \dots, \nabla g_q\left(u^*\right)\right] = q, \\ \nabla g_i\left(u^*\right) = \left(\partial g_i\left(u^*\right)/\partial u_1,\ \partial g_i\left(u^*\right)/\partial u_2, \dots,\ \partial g_i\left(u^*\right)/\partial u_n\right), \\ (i = 1, 2, \dots, q). \end{gathered} \tag{10}$$

Then, if $\tilde{u}_{j_l}$ $(l = q+1, q+2, \dots, n)$ there is an optimal solution to the generated auxiliary variational problem of dimension r=n-q, then the optimal solution to the original prediction problem of dimension n is determined by the following relations:

$$u_{j_i}^{*} = \varphi_i\left(\tilde{u}_1, \tilde{u}_2, ..., \tilde{u}_{j_1-1}, \tilde{u}_{j_1+1}, ..., \tilde{u}_{j_q-1}, \tilde{u}_{j_q+1}, ..., \tilde{u}_{n-1}, \tilde{u}_n\right),$$

$$(i = 1,2,...,q), \quad (11)$$

$$u_l^{*} = \tilde{u}_l, \; l \neq j_i, i = 1,2,...,q.$$

The system of functional relations (5), which appears in the formulation of the statement, reflects the necessary and sufficient conditions of the extremum in the considered forecasting problem.

4. Comparative computational experiments to evaluate the effectiveness of the forecasting methodology, in the framework of a refined variational formulation based on the application of necessary and sufficient extremum conditions to identify promising areas for finding the optimal solution

In the framework of refined variational formulations of inverse forecasting problems, comparative computational experiments were conducted to evaluate the effectiveness of the developed multiparametric models for predicting optimal complexity, which are a generalization of the Arrhenius equations. We denote the defining property of the material by R (residual resource durability reliability, etc.), and its initial value by R_0. We denote the total number of factors affecting the material, and leading to its damage and premature aging through p. Denote by $F_j\left(u_{j,1}, u_{j,2}, ..., u_{j,l_j}; t\right), (j = 1,2,..., p)$ the model describing the effect of the j-th factor on the determining property of the material, $u_{j,1}, u_{j,2}, ..., u_{j,l_j}$ is a system of undefined model parameters, l_j is the number of undefined model parameters.

Each of the functions F_j , describing the effect of the j-th factor on a complex composite, can be represented as a series expansion in a certain system of basic functions $\psi_{kj}\left(\beta_{kj}; t\right), (k = 1,2,3,...)$, which most fully characterize the features of the process of increasing the damage of the material under the influence of extreme environmental factors. Choosing the system of exponential functions describing the nature of processes occurring at the molecular level in accordance with the Arrhenius equation as a system of reference functions when presenting descriptions of factors F_j affecting polymer and composite materials in the form of series, the following universal forecasting model is constructed, describing at the physical level the processes occurring in composite materials and structures under the simultaneous influence of several destabilizing physical factors:

$$R(\,;\,t) = R_0 + \sum_{j=1}^{p}\sum_{k=1}^{N_j} \alpha_{kj}\left(u_{j,1}, ..., u_{j,l_j}\right)\psi_{kj}\left[\beta_{kj}\left(u_{j,1}, ..., u_{j,l_j}\right); t\right]. \quad (12)$$

In these notations: $\alpha_{kj}(u_{j,1}, ..., u_{j,l_j}), \beta_{kj}(u_{j,1}, ..., u_{j,l_j}), \; (j = 1,2,..., p; \; k = 0,1,2,...)$ undefined model parameters describing the impact of the j-th factor. The value of N_j corresponds to the number of terms when the j-th factor is decomposed into a series by the system of basic functions, in particular by the system of exponential functions, if the corresponding physical factor Fj describes the aging processes.

The case of two factors (p=2) describing the aging and hardening processes is considered.
The combined effect of two factors characterizing the aging and hardening processes on a polymer fiber composite (PFC) was studied. To construct an optimal model for predicting optimal complexity,

6 experimental measurements of the residual life of PFC were used in the retrospection interval of 6 years (m=6). The measurements were carried out by year: t=1, 2,..., 6 years. The results of the forecast based on the developed methodology of the optimal model for predicting optimal complexity are shown in Figure 1. The optimal number of model parameters is n*=8.

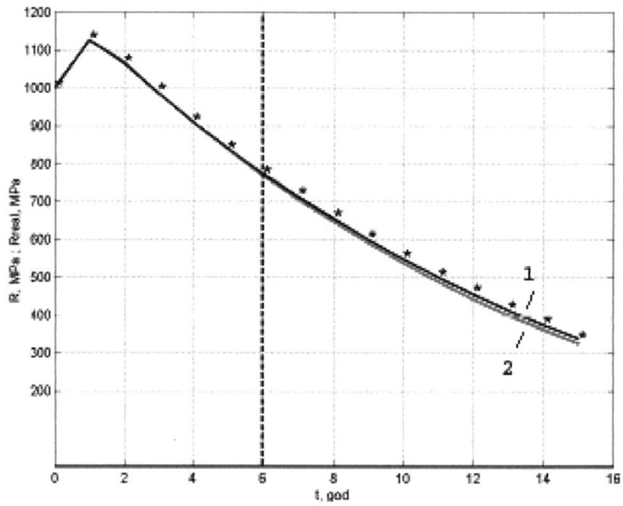

Figure 1. Application of optimal models for predicting optimal performance difficulties for solving problems of high-precision forecasting. Number of experiments data m=6; 1 - real dependence of the residual resource; 2-predicted dependence of the residual resource.

Table 1 shows the results of forecasting based on the optimal eight-parameter exponential model with the number of experimental data m=6. R_t^* - the real dependence of the residual resource; $R(u^{8^*};t_i)$ - the predicted dependence of the residual resource.

Table 1. Prediction results based on the optimal eight-parameter model exponential model for m=6

t, years	R_t^*, MPa	$R(u^{8^*};t_i)$, MPa	ε_t, MPa	ε_t,%
7	712.03	705.18	-6.85	0.96
8	654.03	645.63	-8.40	1.28
9	599.68	589.98	-9.70	1.62
10	548.76	537.00	-11.76	2.14
11	501.04	489.43	-11.61	2,32
12	456.31	444.06	-12.25	2.68
13	414.38	401.67	-12.71	3.07
14	375.09	362.06	-13.03	3.47
15	338.25	325.06	-13.19	3.90

Thus , when the number of experimental data m=6 is considered, when the number of experimental data m is slightly less than the number of parameters n of the eight-parameter model (by 25-30%), reliable forecasting results can be obtained for a time interval of the order of several years (the forecast error is up to 4% when the forecast is up to 9 years). In particular, for the next 5 years, the forecast error does not exceed 2.5 %; for the next 7 years, the forecast error does not exceed 3.1 % ; for the next 9 years, the forecast error does not exceed 4 %,

5. Conclusion

Promising ways to increase the potential capabilities associated with solving problems of predicting the residual life, reliability, and durability of polymer composites under the influence of extreme environmental factors and operational loads are analyzed . The comparative analysis of existing approaches and their possible modifications allowed us to conclude that the most promising opportunities are associated with the development of a methodology for effectively narrowing the search areas containing globally optimal solutions. According to the developed methods, based on the application of necessary and sufficient extremum conditions for the effective constructive transformation of variational statements of inverse problems of predicting the residual resource of composites to variational problems of lower dimension, comparative computational experiments were carried out, which showed the effectiveness of the developed approach. Computational experiments have shown the high efficiency of the developed approach for constructing optimal forecasting models of optimal complexity, based on the immersion of the original forecasting problem in a parametric family of problems , in which the parameter is the number of undefined parameters of the model. Computational experiments have shown high efficiency of the developed approach for constructing optimal forecasting models of optimal complexity, based on the immersion of the original forecasting problem in a parametric family of problems, in which the parameter is the number of indeterminate parameters of the model.

Acknowledgments
This work was performed as part of a state assignment, state registration number AAAA-A21-121011590012-9 , AAAA-A19-119012290177-0

References

[1] Stepanov M.N., Zimin A.V. Prediction of fatigue resistance characteristics of materials and structural elements. 2007. Moscow. Innovative mechanical engineering. 392 p. (*in Russian*)
[2] Lurie S. A., Solyaev Yu. O., Nguen D. Q. 2015. *An International Journal.Composites: Mechanics, Computations, Applications* **6** 279 - 291
[3] Gusev E. L. Mathematical methods for the synthesis of layered structures. 1993. Novosibirsk. Nauka. 262 p. (*in Russian*)
[4] Gusev E. L. 1996 . *Reports of the Academy of Sciences.* **346** 324 - 326
[5] Gusev E. L., Bakulin V. N. 2016. *Journal of Engineering Physics and Thermophysics* **89** 260-264
[6] Gusev E. L., Bakulin V. N. 2018 . *Doklady physics.* **63** 213- 217 doi: 10.1134/S1028335818050075.
[7] Gusev E. L., Bakulin V. N. 2018 . *Doklady physics* **63** 388- 392 doi: 10.1134/S1028335818090082
[8] Reutov A.I. Forecasting the reliability of construction products made of polymer materials. 2007. Moscow. Stroymaterialy. 184 p. (*in Russian*).
[9] Potapova L.D., Yartsev V.P. 2005. Mechanics of materials under complex stress state : how to predict the limit values stress? Moscow: Mashinostroenie. 242 p. (*in Russian*).
[10] Reutov A.I., Sidorenko A.A., Filatov I.S. 2002. *Proceedings of Tomsk Polytechnic University.* **305** 96-203. (*in Russian*).
[11] Reutov A.I. 2009. *Bulletin of the Tomsk state university of architecture and construction.* **2**

127-141. (*in Russian*).
[12] Gusev E. L. 2016. *Material Physics and Mechanics* **26** 70 -72
[13] Gusev E. L., Bakulin V. N., Chernykh V. D. 2019. *IOP Conf. Series: Journal of Physics:Conf. Series* 1392 (2019) 012008. IOP Publishing doi: 10.1088/1742-6596/1392/1/012008 .
[14] Gusev E.L. , Babenko F.I. 2015. Collection of works «Materials for technical devices and structures used in the Arctic». Moscow. All-Russian Institute of Aviation Ma terials 132-141. (*in Russian*).
[15] Gusev E.L. 2015. *Journal of Machinery Manufacture and Reliability.* **44** 148–152.
[16] Gusev E. L., Bakulin V. N. 2015. *Mechanics of composite materials* **51** 637 -644. doi: 10.1007/s11029-015-9533-4

Mathematical modeling of the VHCF resonance loadings with a progressive damage accumulation

I S Nikitin, A D Nikitin and B A Stratula

Institute for computer aided design of RAS, 2nd Brestskaya st., 123056, Moscow, Russia

e-mail: i_nikitin@list.ru

Abstract. One of the most critical in-situ loading modes is resonant. Such loading conditions may lead to premature fatigue damage accumulation and failure. The progressive fatigue damage changes the natural frequency of a structure and causes a permanent modification of loading conditions. This paper considers the model of fatigue damage accumulation regarding its evolution and resonant loading parameters. To describe the evolution of material degradation due to fatigue a complex kinetic model of damage development and numerical algorithm are introduced. The model contains the evolution equation for the damage function depending on stress state. The algorithm is based on a special post proceeding of a numerical solution for an elastic solid mechanics problem that can be also applied for elastic-plastic case. Different loading conditions can be considered by choosing the method of solution for solid mechanics problem. This paper is focused on the comparison between quasi-static and resonant loading conditions used for damage function calculation. It was shown that use of resonant loading solution leads an earlier crack initiation and shorter fatigue life results compared to quasi-static solution. A spatial development of the fatigue damage zone is also different for these two cases. The normal crack opening mechanism is discussing in this paper.

1. Introduction

It is well known that fatigue failure is time consuming process and related to the degradation of the material versus cycles. However, for the engineering application is it usually used the criteria that do not consider the degradation process. In last few decades, several classes of criteria have been constructed that predict the number of cycles before the initiation of fatigue damage (micro-cracks) based on the stress amplitudes. These stress amplitudes or maximum values in the cycle (or average) are estimated at the first load step and characterize the uniform stress-strain state of the working part of the specimen. This class can be described as stress-based criterion. Many of stress criteria are based on a direct generalization of the S-N Wöhler-type curves described by Basquin-type relations [1] for fatigue tests. Reviews on this topic are given in [2–4].

However, the process of material degradation is important because it may lead to redistribution of stress field or local stress relaxation. In this paper we study the processes of fatigue damage zones development using the damage theory approach dating back to [5,6]. It is a sort of generalization of the Kachanov - Rabotnov damage theory for the case of cyclic loading. In the application to the cyclic loading and fatigue failure problems this approach was applied in [7,8]. A complex model for the development of fatigue failure with use of the evolutionary equation for the damage function is proposed. The model is associated with two different fatigue criteria. These criteria describe different micro-crack development types. The model parameters are determined for various fatigue failure modes – low-cycle and high-cycle fatigue (LCF, HCF).

The following description of the whole SN-curve is used. Up to number of cycles $N \sim 10^3$ there is the regime of repeated-static loading with amplitude slightly differing from the static tensile strength σ_B is realized. Further the fatigue curve (Wöhler curve) describes the modes of the LCF-HCF up to $N \sim 10^7$ cycles with an asymptotic approach to the fatigue limit σ_u.

It should be noted that at present there is an idea of an explicit division of the classic Wöhler branch into two parts – in fact, the LCF and the HCF. The boundary of this transition region is determined not by the value of *N*, but by the loading amplitude that is equal to the yield strength of the material σ_T [9], since this means a change of the physical mechanism of fatigue failure. In addition, the boundary of the repeated-static range $N \sim 10^3$ is rather arbitrary. It is also specified in [9] depending on the strength and plastic characteristics of the material. However, in this paper we will keep the suggestion of the proposed model of damage development based on the scheme described above.

The model associated with the two well-known criteria for multiaxial fatigue failure is proposed. These criteria imply different micro-crack development types. The first one describes the fatigue failure associated with the tensile micro-cracks development. It is the stress-based Smith–Watson–Topper criterion [10] described in [11], in which the amplitudes of maximum tensile stress play a decisive role in the development of fatigue damage.

For numerical simulation, a simple hourglass specimen is used.

2. Very high cycle experimental procedure

The study on the very-high cycle fatigue (VHCF) properties of the structural materials have been performed by using high-frequency fatigue testing system. The principle of such machine operation is based on the piezoelectric effect when the electric signal converts to mechanical vibrations of the same frequency. The system is operating in the resonance regime at the natural frequency of the first harmonic of the longitudinal elastic wave. The fatigue testing system, fig.1, consists of personal computer with the communication card, generator, converter, booster, ultrasonic horn, specimen, and cooling system. The special software controls the amplitude and frequency of the loading via high-performance feedback. The algorithm of the software keeps the constant loading amplitude and adjust the loading frequency to the natural frequency of the system. The cooling system is used for keeping the piezoelectric crystals and VHCF specimen temperature at the standard conditions (about the room temperature). An additional infrared camera procedure was engaged to verify the temperature at the specimen surface during the VHCF loading. The result shows a slight temperature elevation during the tests (about 15 – 20 degree Celsius). Therefore, the loading conditions were room temperature, laboratory air environment, constant amplitude loading, loading frequency is about 20 kHz.

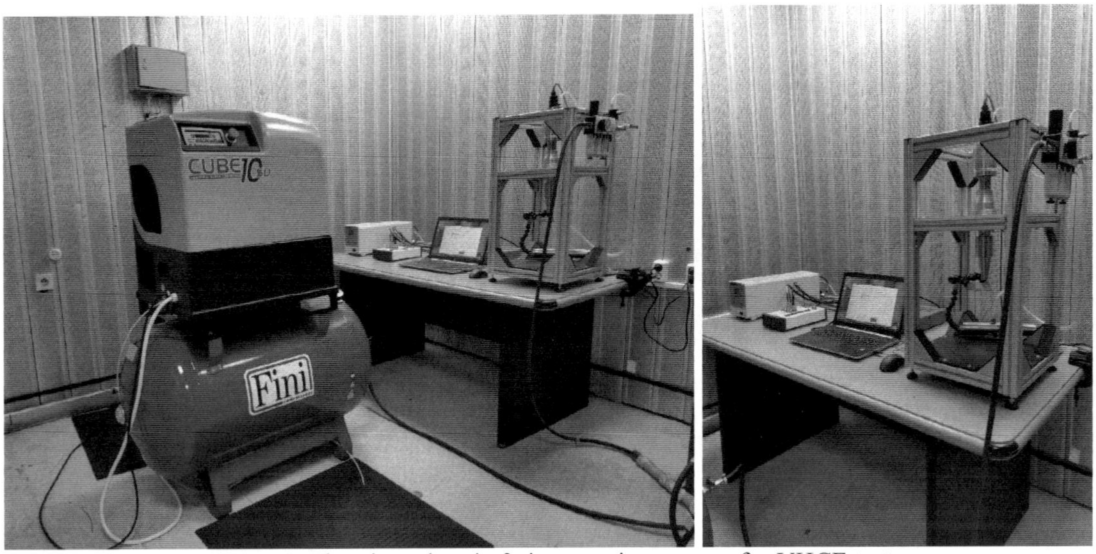

Figure 1. The piezoelectric fatigue testing system for VHCF tests

The fatigue tests were performed by using staircase method (ASTM) with the base of 10^9 cycles and stress amplitude increment of 15 MPa. The specimens for the VHCF tests were design according to gigacycle concept [12] and have the hourglass shape. After the fatigue failure all the fracture surfaces were investigated by using scanning electron microscopy (SEM). The results of observations were compared with the prediction of numerical simulations for damage shape.

3. Very-high cycle concept and analytical solution

The concept of the very-high cycle fatigue is based on the usage of elastic wave energy. It is assumed that elastic wave starting from one side of the specimen propagate through the material without significant energy dispersion and reflected from another end. The incident and reflected waves add up to each other to form a standing wave. This standing wave has a node of vibration in the middle of the specimen. The stationary standing wave can be installed in the specimen at a given frequency for a certain resonance length. For the simple specimen shaped in the form of strait cylinder or bar the resonance length should be equal to the half-length of the incident wave [12]. The resonance length, strain and stress distribution in this case can be found analytically. In the case of the hour-glass specimens the analytical solution is also possible. The analytical solution is based on the solution of the differential equation for elastic wave propagation.

$$\rho\frac{\partial^2 u}{\partial t^2}=\frac{E}{(1+\nu)}\left(\frac{1}{1-2\nu}\frac{\partial e}{\partial x}+\nabla^2 u\right); \quad c_l=\sqrt{\frac{E(1-\nu)}{(1+\nu)(1-2\nu)\rho}} \tag{1}$$

The hour-glass specimen is separated into two calculation part: constant cross section bar and reduced cross-section part. The solutions for these two parts should satisfy to continuity conditions at connection locations and boundary conditions at the specimen's ends. The differential equation for the reduced part will be

$$\frac{\partial^2 u(x,t)}{\partial t^2}=c^2\left\{p(x)\frac{\partial u(x,t)}{\partial x}+\frac{\partial^2 u(x,t)}{\partial x^2}\right\}=0, \text{ where } p(x)=\frac{S'(x)}{S(x)} \tag{2}$$

The solution of such equation can be founded in the following form $u(x,t)=U(x)\cdot sin(\omega t)$. Substitution of this solution to the differential equation will give us

$$U''(x)+p(x)U'(x)+k^2U(x)=0, \text{ where } k=\omega/c \tag{3}$$

In order to get the analytical solution for this equation the reduced section should be reproduced by a harmonic function. According to [12] the hyperbolic cos feet good the circular profile of the reduced gage section. In this case the analytical solution for the reduced part can be obtained in the following form

$$U(x)=A_0\phi(L_1,L_2)\frac{\sin(\beta\cdot x)}{\cosh(\alpha\cdot x)},\ |x|\le L_2, \text{ where } \phi(L_1,L_2)=\frac{\cos(k\cdot L_1)\cosh(\alpha\cdot L_2)}{\sinh(\beta\cdot L_2)} \tag{4}$$

and for the constant cross section part of the specimen the analytical solution will be

$$U(x)=A_0\cos\left[k(L-x)\right],\ L_2\le|x|\le L \tag{5}$$

Based on the obtained analytical solution for the reduced section the strain and stress distributions can be obtained in the following form

$$\varepsilon(x)=A_0\phi(L_1,L_2)\frac{\left[\beta\cosh(\beta\cdot x)\right]\cosh(\alpha\cdot x)-\alpha\sinh(\beta\cdot x)\sinh(\alpha\cdot x)}{\cosh^2(\alpha\cdot x)} \tag{6}$$

$$\sigma(x)=E_dA_0\phi(L_1,L_2)\frac{\left[\beta\cosh(\beta\cdot x)\right]\cosh(\alpha\cdot x)-\alpha\sinh(\beta\cdot x)\sinh(\alpha\cdot x)}{\cosh^2(\alpha\cdot x)} \tag{7}$$

Where A_0 is an amplitude of applied displacements, E_d is dynamic Young's modulus. Using the equation 1 the stress amplitude in the middle of specimen (x=0) can be obtained as function of external load. This analytical solution can be used to verify the numerical solution of resonant problem.

4. Kinetic equation for damage in LCF-HCF and VHCF mode

Various criteria use different stress combinations to calculate an equivalent stress value. Some of them are based on normal stress components of a stress state while other based on shear components. In this paper we are going to discuss only the normal crack opening mechanism associated with the stress-based Smith–Watson–Topper criterion. The considered model of damage evolution in the case of cyclic loads is partially presented in [13] for the description of damages during dynamic loading. The generalization of this model for different modes or regimes of fatigue is presented in [14, 15]. This generalized model assumed the normal crack opening as a driven mechanism of fatigue destruction. The application of this model for fatigue life and damage zone prediction in the hourglass specimens subjected to VHCF loading is presented here.

4.1. Generic equations

At first, let us introduce some variables that are used later.

We can write a generic fatigue fraction criterion corresponding to the left branch of the bimodal fatigue curve in the following form:

$$\sigma_{eq}=\sigma_u+\sigma_L N^{-\beta} \tag{8}$$

From the condition of repeated-static fracture up to values of $N\sim10^3$ by the method [4] it is possible to obtain the value $\sigma_L=10^{3\beta}(\sigma_B-\sigma_u)$. In these formulas σ_B is the static tensile strength of the material, σ_u is the classic fatigue limit of the material during a reverse cycle (asymmetry coefficient of the cycle R = -1), β is power index of the left branch of the bimodal fatigue curve.

From the fatigue fracture criterion, we obtain the number of cycles before fracture at uniform stressed state:

$$N = 10^3 \left[(\sigma_B - \sigma_u) / \langle \sigma_{eq} - \sigma_u \rangle \right]^{1/\beta} \tag{9}$$

In order to describe the process of fatigue damage development in the LCF-HCF mode, a damage function $0 \le \psi(N) \le 1$ is introduced, which describes the process of gradual cyclic material failure. When $\psi = 1$ a material particle is considered completely destroyed. Its Lame modules become equal to zero. The damage function ψ as a function on the number of loading cycles for the LCF-HCF mode is described by the kinetic equation:

$$\partial \psi / \partial N = B \psi^{\gamma} / (1 - \psi^{\alpha}) \tag{10}$$

where α and $0 < \gamma < 1$ are the model parameters that determine the rate of fatigue damage development. The choice of the denominator in this two-parameter equation, which sets the infinitely large growth rate of the zone of complete failure at $\psi \to 1$, is determined by the known experimental data on the kinetic growth curves of fatigue cracks, which have a vertical asymptote and reflects the fact of their explosive, uncontrolled growth at the last stage of macro fracture.

An equation for damage of a similar type was previously considered in [8], its numerous parameters and coefficients were determined indirectly from the results of uniaxial fatigue tests. In our case, the coefficient B is determined by the procedure that is clearly associated with the selected criterion for multiaxial fatigue failure of one type or another. It has the following form.

The number of cycles to complete failure N at $\psi = 1$ is found from the equation for damage for a uniform stress state:

$$\left[\psi^{1-\gamma} / (1-\gamma) - \psi^{(1+\alpha-\gamma)} / (1+\alpha-\gamma) \right]\Big|_0^1 = B N\Big|_0^N, \quad N = \alpha / (1+\alpha-\gamma) / (1-\gamma) / B \tag{11}$$

By equating the values N from the fracture criterion (9) and from the solution of the equation for damage (11), we obtain the expression for the coefficient B:

$$B = 10^{-3} \left[\langle \sigma_{eq} - \sigma_u \rangle / (\sigma_B - \sigma_u) \right]^{1/\beta} \alpha / (1+\alpha-\gamma) / (1-\gamma) \tag{12}$$

where the value σ_{eq} is determined by the selected mechanism of fatigue failure and the corresponding multiaxial criterion (8).

Similarly, for the VHCF regime (the right branch of the bimodal SN curve)

$$\sigma_L = 10^{8\beta} (\sigma_u - \tilde{\sigma}_u)$$

where $\tilde{\sigma}_u$ – fatigue limit of the material under symmetric loading cycle in the VHCF regime.

The expression for the coefficient B will be following [15]:

$$B = 10^{-8} \left[\langle \sigma_{eq} - \tilde{\sigma}_u \rangle / (\sigma_u - \tilde{\sigma}_u) \right]^{1/\beta} \alpha / (1+\alpha-\gamma) / (1-\gamma)$$

Finally [15], for the LCF-HCF at $\sigma_u + \Delta\sigma_u < \sigma_{eq} < \sigma_B$, $\Delta\sigma = 10^{-5\beta} (\sigma_B - \sigma_u)$

$$B = 10^{-3} \left[\langle \sigma_{eq} - \sigma_u \rangle / (\sigma_B - \sigma_u) \right]^{1/\beta} \alpha / (1+\alpha-\gamma) / (1-\gamma)$$

And for VHCF at $\tilde{\sigma}_u < \sigma_{eq} \le \sigma_u + \Delta\sigma_u$

$$B = 10^{-8} \left[\langle \sigma_{eq} - \tilde{\sigma}_u \rangle / (\sigma_u - \tilde{\sigma}_u) \right]^{1/\beta} \alpha / (1+\alpha-\gamma) / (1-\gamma)$$

when $\sigma_{eq} \le \tilde{\sigma}_u$ the fatigue failure does not occur, when $\sigma_{eq} \ge \sigma_B$ the failure comes instantly.

4.2. *Smith–Watson–Topper criterion*

The criterion of multiaxial fatigue failure in the LCF-HCF mode with the development of normal-stress micro-cracks (stress-based SWT) corresponding to the left branch of the bimodal fatigue curve has the form:

$$\sqrt{\langle \sigma_{1_{\max}} \rangle \Delta\sigma_1 / 2} = \sigma_u + \sigma_L N^{-\beta} \tag{13}$$

where σ_1 is the largest principal stress, $\Delta\sigma_1$ is the spread of the largest principal stress per cycle, $\Delta\sigma_1 / 2$ is its amplitude. According to the chosen criterion only tensile stresses lead to failure, so it has the value $\langle \sigma_{1_{\max}} \rangle = \sigma_{1_{\max}} H(\sigma_{1_{\max}})$. We assume that the right branch of SN-curve for the VHCF regime is also describing by the equation (13).

Such, the equivalent stress has the following expression:

$$\sigma_{eq} = \sqrt{\langle \sigma_{1_{\max}} \rangle \Delta\sigma_1 / 2} \tag{14}$$

5. Fatigue damage development calculation algorithm

The software realizing the finite element methods was used to calculate the loading cycle of a deformable specimen, supplemented by a code to calculate the damage equation and changes of elasticity modulus.

5.1. *General approach to damage function*

To integrate the equation $d\psi / dN = B\psi^{\gamma} / (1-\psi^{\alpha})$, the damage function approximation was applied at the *k*-node of the computational grid for given discrete values ψ_k^t at moments N^t and sought ψ_k^{t+1} at moments N^{t+1}.

To calculate the damage equation, the value $\alpha = 1-\gamma$ was chosen for which by analytic integration an explicit expression for $\psi_k^{t+1}(\psi_k^t, \Delta N^t)$ can be obtained:

$$\left[\psi^{1-\gamma} / (1-\gamma) - \psi^{2(1-\gamma)} / 2 / (1-\gamma)\right]\Big|_{\psi_k^t}^{\psi_k^{t+1}} = B N\Big|_{N^t}^{N^{t+1}} \tag{15}$$

With the denotations $(\psi_k^{t+1})^{1-\gamma} = x$, $q = 2(1-\gamma)B\Delta N^t + (\psi_k^t)^{1-\gamma} - 2(\psi_k^t)^{2(1-\gamma)}$ and $\Delta N^t = N^{t+1} - N^t$ the equation transforms to $x^2 - 2x + q = 0$ and its valid root $x = 1 - \sqrt{1-q} < 1$.The damage parameter depends on the increment of the number of cycles ΔN^t as:

$$\psi_k^{t+1} = \left(1 - \sqrt{1 - \left[2(1-\gamma)B\Delta N^t + (\psi_k^t)^{1-\gamma} - 2(\psi_k^t)^{2(1-\gamma)}\right]}\right)^{1/(1-\gamma)} \tag{16}$$

Here increment value ΔN^t defined as follows. Based on the stress state calculation data, the coefficient B is calculated for each node. After that, for each node, the following values are calculated

$$\Delta\tilde{N}_k^t = \left[\psi^{1-\gamma} / (1-\gamma) - \psi^{2(1-\gamma)} / 2 / (1-\gamma)\right]\Big|_{\psi_k^t}^{1} / B \tag{17}$$

corresponding to the number of cycles at which in the node k from its current level of damage and equivalent stress complete destruction will be achieved (damage is equal to 1).If the damage level in the considered node is less than the threshold ψ_0 (threshold $\psi_0 = 0.95$ is selected), then the value for this node $\Delta\tilde{N}_k^t$ is multiplied by a factor of 0.5. Otherwise, it is multiplied by a factor of 1. Thus, the step of incrementing the number of cycles for a given node is $\Delta N_k^t = 0.5(1 + H(\psi_k^t - 0.95))\Delta\tilde{N}_k^t$. Of all the ΔN_k^t values the smallest one is selected. The increment of the number of loading cycles for the calculation of the entire specimen is $\Delta N^t = \min_k \Delta N_k^t$. For each node, based on its current level of damage and equivalent voltage, a new level of damage is found taking into account the calculated increment ΔN^t.

5.2. Material properties change

All elements are sorted out, for each of them the most damaged node is searched and according to its damage the mechanical properties of the element are adjusted:

$$\lambda(\psi_k^t) = \lambda_0(1 - \kappa\psi_k^t),\ \mu(\psi_k^t) = \mu_0(1 - \kappa\psi_k^t) \tag{18}$$

Those elements that belong to nodes with damage $\psi = 1$ are removed from the calculation area and form a localized zone (crack-like) of completely destroyed material. The calculation ends when the boundaries of a completely damaged region exit to the specimen surface (macro destruction) or the evolution of this region stops.

6. Calculation results

We performed a series of experiments to determine the effect of the proposed scheme on fatigue behavior of a specimen. We wanted to study an effect of different loading regimes: quasi static and resonant on fatigue life and spatial distribution of damage zone. Both regimes were studied under uniaxial tensile loading.

The Ansys software was used to solve the quasi static and resonant solid mechanics problems. The obtained stress distribution is used to calculate the progressive fatigue damage for a loading cycle of a deformable specimen. For this purpose, the finite element software was supplemented by a code to calculate the damage equation and changes of elasticity modulus.

6.1. Initial conditions

All tests were conducted on an hourglass specimen. The shape and dimensions of the studied specimen are presented on figure 2-a and the mesh on figure 2-b.

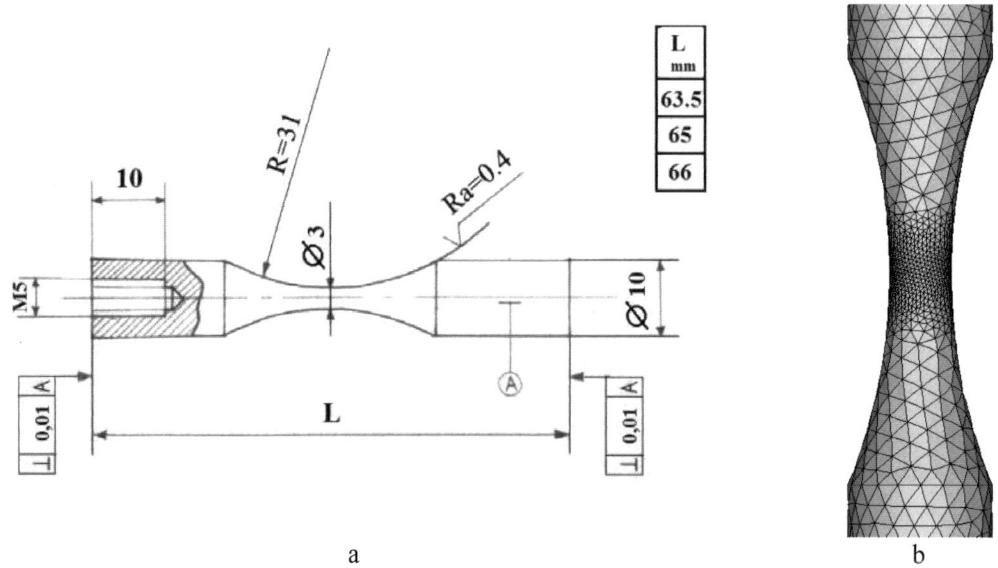

Figure 2. Shape, dimensions, and mesh of the specimen.

The loading parameter is displacement applied to the end(s) of the specimen for both quasi-static and resonant problems. In the case of quasi-static problem, the displacements are applied to both ends of the specimen. For the resonant calculations, the modal analysis was performed before the harmonic loading. The displacements were applied to one of the specimen bottoms. In all numerical experiments all loads were in the same phase and with asymmetry ratios $R = -1$.

The material used in calculation is titanium alloy with the following ultimate tensile strength and fatigue parameters $\sigma_B = 1135$ MPa, $\sigma_u = 330$ MPa, $\tilde{\sigma}_u = 250$ MPa, $\beta = 0.3$ (HCF), $\beta = 0.25$ (VHCF). Elasticity modulus of intact alloy are $\lambda_0 = 77$ GPa, $\mu_0 = 44$ GPa.

6.2. *Numerical experiments*

6.2.1. *Quasi-static loadings*

The series of numerical simulations on fatigue damage accumulation and crack like defect growth in hour-glass specimen were performed based on the quasi-static solution. The damage function was calculated based on the elastic solution obtained for the specimen loaded at both ends by displacements. The values of the external load were chosen in the wide range that allow us to engage different crack initiation mechanisms from high- to very-high cycle fatigue. The figure 3 shows the result of calculation when the external load equal to 40 micrometers that corresponds to HCF range. According to the numerical simulation the estimated fatigue life is 1.94×10^6 cycles. The normal stress distribution, early crack like defect initiation and developed zone of fatigue damage is shown on Fig.3.

The fatigue damage zone is located at the specimen surface that is in good agreement with experimental results for HCF.

IOP Conf. Series: Materials Science and Engineering 1191 (2021) 012023 doi:10.1088/1757-899X/1191/1/012023

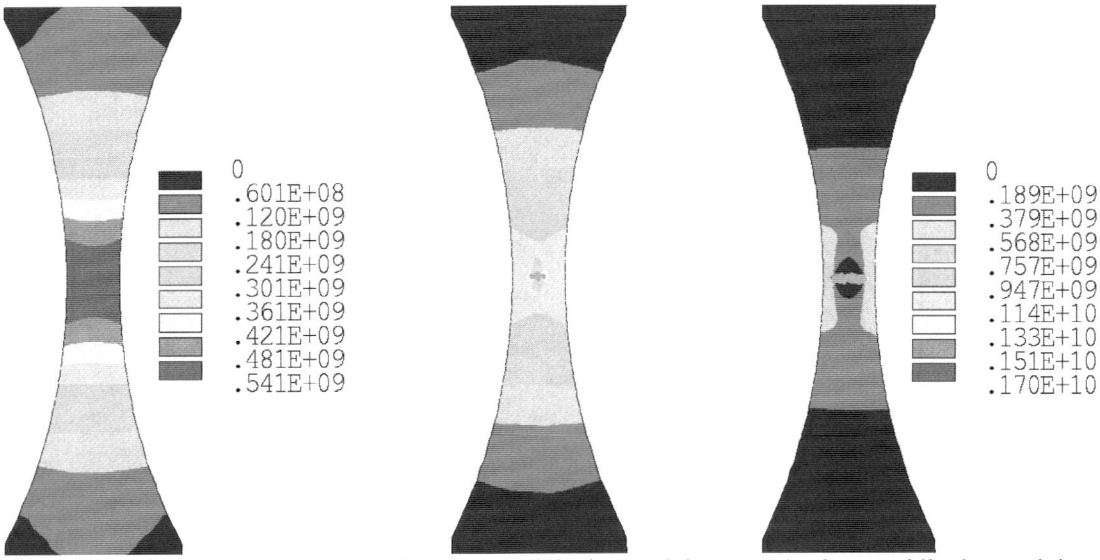

a) Stress at the crack nucleation b) Stress while the crack is growing c) Stress while the crack is growing

Figure 3. Normal-type crack growth; a) and b) images are shown in the cross-section from, c) is from the top. Crack-like zone is colored with gray.

The numerical simulation in the range of very-high cycle fatigue were performed under corresponding external load of 21 micrometers. The results of calculation are given on the Fig.4. The estimated fatigue life for this case is 8.23×10^9 cycles. The initial rate of damage function accumulation was higher for the surface layers of the specimens, however at later stages the internal fatigue damage accumulation turns to be more intense. The final crack like defect developed from the subsurface that is also typical for VHCF loading.

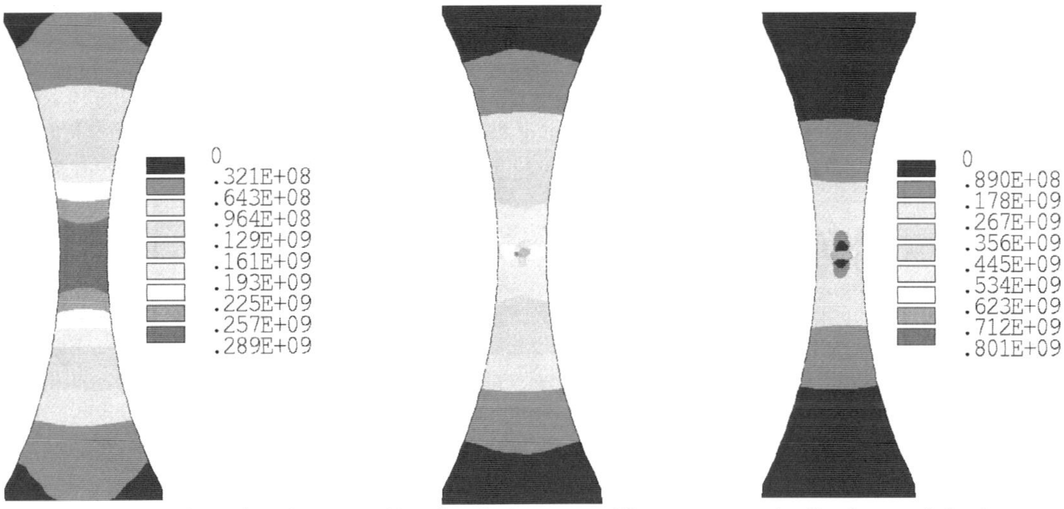

a) Crack nucleated at the surface, cross-section view b) Crack then grew like a normal-type crack, cross-section view c) Surface path is shown here.View of the outer surface of the specimen.

Figure 4. Crack nucleated at the surface due to shear components of the stress.

IOP Conf. Series: Materials Science and Engineering 1191 (2021) 012023 doi:10.1088/1757-899X/1191/1/012023

The numerical simulation of the fatigue damage accumulation shows a good agreement with the experimental data and fatigue behavior of structural materials under different loading regimes.

6.2.2. Resonant loadings
The numerical simulation of damage function based on resonant problem solution is presented here. Similar to quasi-static case the results of resonant method are obtained first for the HCF range Fig.5 and after for the VHCF, fig.6. The applied amplitude is 35 micrometers and corresponding fatigue life $N_f=1.11\text{x}10^6$ cycles. The spatial location of the damage zone is different from obtained in quasi-static.

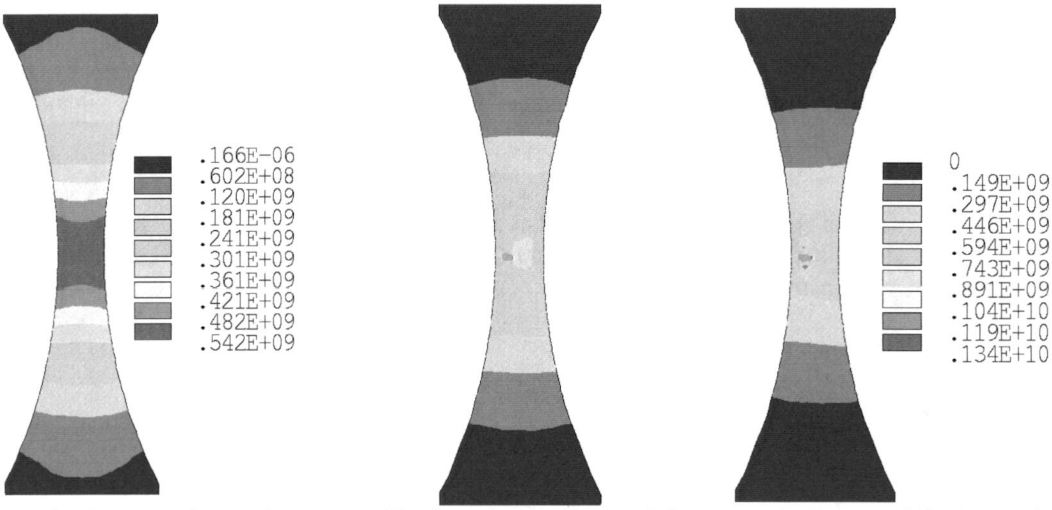

a) Stress at the crack nucleation b) Stress while the crack is growing c) Stress while the crack is growing

Figure 5. Normal-type crack growth; a) and b) images are shown in the cross-section from, c) is from the top. Crack-like zone is colored with gray.

The fatigue life of the specimen under resonant loading is shorter even under lower external load. Such the resonant method of calculation provides more safe design compared to quasi-static. The same analysis for the case of VHCF loading, external load is 19 micrometers and corresponding fatigue life is $2.94\text{x}10^9$ cycles is shorter, compare to the quasi-static estimation. This can be explained by lower external load. Nonetheless, the following tendency can be found: in the HCF range the results of numerical simulation are more sensitive to the resonant method compared to the case of VHCF.

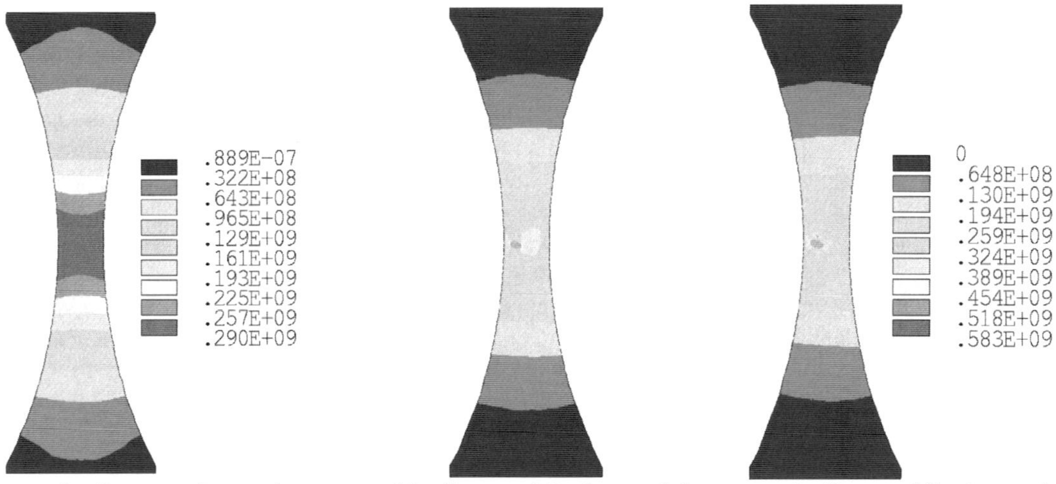

a) Stress at the crack nucleation
b) Stress while the crack is growing
c) Stress while the crack is growing

Figure 6. Normal-type crack growth; a) and b) images are shown in the cross-section from, c) is from the top. Crack-like zone is colored with gray.

The resonant problem is more complex compare to quasi-static one, because it should be solved in several steps. At the first step the modal analysis for getting natural frequencies of the specimens should be performed. At the second step the harmonic analysis at a given frequency should be performed. As a result of these calculations, we will get a database of stresses for each node of the mesh. To post-proceed these data the same to quasi static case algorithm is engaged. The difficulties of this solution are related to permanent decreasing of the specimen natural frequency with degradation of the material. Therefore, the modal analysis step should be repeated before each iteration of damage function calculation because of material degradation. The results of frequency drop versus progressive damage (number of cycles) is presented on Fig.7.

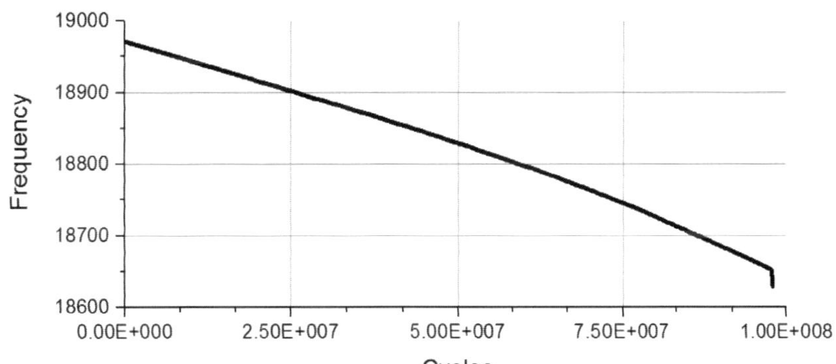

Figure 7. The natural frequency evolution versus number of cycles corresponding to progressive damage accumulation

6.2.3. Comparison quasi-static and resonant loadings

The results of both methods (based on quasi-static and resonant) show a good qualitative agreement with experimental data. The first method developed in [15] show fairly good location of the fatigue crack in the middle of the specimen for HCF and VHCF ranges. The second method shows scatter of spatial location of fatigue damage zone that is also physical behavior of the material. In the case of numerical simulation, it means that resonant method is more sensitive to the mesh quality and permanent changing of the natural frequency leads to higher perturbation of damage zone growth. In the case of quasi-static method, the loading condition is less depending on progressive damage and the spatial scatter is lower. An interesting feature is higher sensitivity of fatigue life estimation from the used method for the range of HCF while in the case of VHCF loading the effect of used method is almost absent. The result of calculation for HCF range shows almost the same fatigue life under higher external load (N_f=1.94x10^6 at 40 micron) versus resonant case (N_f=1.11x10^6 at 35 micron). In the case of VHCF calculation the results follow the same SN curve: quasi-static method (N_f=8.3x10^9 at 21 micron) versus resonant method (N_f=2.9x10^9 at 19 micron). Beside this difference both methods give a good prediction of spatial location of fatigue damage zone. The close results between quasi-static and resonant method are not surprising because of characteristic time of loading process. In the numerical scheme the two main parameters can be outlined: the rate of the process and the linear size of calculated area. In the case of VHCF loading the rate of the process is related to the elastic wave propagation that is above 5700 m/sec for titanium alloy. The loading frequency is 20 kHz. The typical linear size of the specimen is about 65 mm. Taking all these parameters into account it can be concluded that this resonant regime is closer to quasi-static problem than to dynamic one. Therefore, the obtained small difference between two methods in the VHCF range is normal.

7. Conclusions

The model of complex multi modes fatigue behavior was applied to predict the fatigue life and damage zone location of the hour-glass specimens. The two different methods for solving the corresponding solid mechanics problem were engaged. The first method is based on quasi-static solution and the second one is based on the resonant problem solution. It was found that resonant method provides more safe fatigue design for the HCF range, while in the VHCF both methods are equivalent. For both methods, the crack like defect was observed at the specimen's surface under HCF loading (nope, it was also under the surface) and at the subsurface in the case of VHCF. The resonant method is more sensitive to the progressive material degradation and lead to higher perturbation of crack like path development. In the case of quasi-static method, the crack-like defect is always located in the middle section of the specimen. The resonant methods allow to consider the stiffness drop due to material degradation that is new step of model development.

Acknowledgments
The study was carried out with a grant from the Russian Science Foundation (project No. 19-19-00705).

References

[1] Basquin O H 1910 The exponential law of endurance tests*Proc of the American society for testing and material* **10** 625–30
[2] Meggiolaro M A, Miranda A C, de Castro J 2007 Comparison among fatigue life prediction methods and stress-strain models under multiaxial loading*Proceedings of 19th Int. Congress of Mech. Eng*
[3] Karolczuk A Papuga J Palin-Luc T 2020 Progress in fatigue life calculation by implementing life-dependent material parameters in multiaxial fatigue criteria*Int. J. of fatigue* **134** 1-13
[4] Bourago NG Zhuravlev AB Nikitin IS 2011 Models of multiaxial fatigue fracture and service life estimation of structural elements*Mechanics of Solids* **46** 828–38
[5] Kachanov L M 1958 O vremeni razrushenija v usloviyah polzuchesty*Izv. AN SSSR OTN***8** 26–31
[6] Rabotnov J N 1959 O mechanizme dlitel'nogo razrushenia. Voprosi prochnosti materialov i konstrukcij *AN SSSR OTN* 5–7
[7] Lemaitre J Chaboche JL 1994 *Mechanics of solid materials* (Cambridge: Cambridge University Press) p 582
[8] Marmi AK Habraken AM Duchene L 2009 Multiaxial fatigue damage modeling at macro scale

of Ti6Al4V alloy*Int. J. of fatigue* **31** 2031–40
[9] Shanyavskiy AA Soldatenkov AP 2019 The fatigue limit of metals as a characteristic of the multimodal fatigue life distribution for structural materials*Procedia Structural Integrity* **23** 63–8
[10] Smith RN Watson P Topper TH 1970 A stress-strain parameter for the fatigue of metals*J. of Materials* **5** 767–78
[11] Gates N Fatemi A 2016 Multiaxial variable amplitude fatigue life analysis including notch effects *Int. J. of fatigue* **91** 337–51
[12] Bathias C Paris P 2004 Gigacycle fatigue in mechanical practice Dekker New-York p.328
[13] Burago NG Nikitin IS Nikitin AD Stratula BA 2019 Algorithms for calculation damage processes *Frattura ed Integrità Strutturale* **49** 212–24
[14] Nikitin IS Burago NG Nikitin AD and Stratula BA 2020 Through calculation method of fatigue damage *IOP Conference Series: Materials Science and Engineering* **927** 927 012019
[15] Nikitin IS Burago NG Zhuravlev AB and Nikitin AD Multi-mode model for fatigue damage development 2020 *Mechanics of Solids* **55** 8 298–306

IOP Conf. Series: Materials Science and Engineering 1191 (2021) 012024 doi:10.1088/1757-899X/1191/1/012024

Free vibrations of liquid in a tank with a partition

Goncharov D.A.[1], Pozhalostin A.A.[1]

[1]Bauman Moscow State Technical University, 2-ya Baumanskaya str. 5, b.1, Moscow 105005, Russia

E-mail: goncharov@bmstu.ru

Abstract. I The article presents an analytical solution to the problem of small oscillations of a liquid in a rigid tank with an elastic baffle. This mechanical system is a model for various oil tankers. The frequency transcendental equation of the problem is obtained in closed form under certain assumptions, the left side of which is a meromorphic function with an infinite number of simple poles. This result is achieved by expanding the homogeneous solution into a series in terms of the eigenfunctions of the Laplace equation in a rectangular coordinate system.

1. Introduction

Global climatic changes call to life various alternative routes for transporting oil and refined products, which were hardly feasible in the past. In particular, it becomes possible for some consumers to transport oil and refined products through the routes of the Northern Sea Route, due to the reduction of the ice cover in the Arctic Ocean. At the same time, one cannot talk about its complete disappearance, the specific features of this route determine a number of specific problems of the dynamic interaction of elastic structures with a fluid. The most obvious example of such an interaction is the impact of a drifting iceberg on the hull of a tanker ship; these problems were considered in [5].

A feature of these interactions is the ability to dynamically involve the entire mass of the transported content, which under certain conditions can lead to catastrophic consequences for the entire vessel. This circumstance gives rise to a number of problems, in particular, the problem of dynamic interaction of an elastic partition with a liquid, the problem of vibrations of an elastic partition in a liquid. Considering this problem, we can introduce a rectangular Cartesian coordinate system, due to the specifics of the division of compartments in oil tankers. Considering the vibrations of an elastic partition in an ideal and incompressible fluid, which well simulates the properties of oil and petroleum products, we can introduce the velocity potential and formulate the continuity condition for the fluid, which in these circumstances will take the form of the Laplace equation in Cartesian coordinates for the velocity potential, which allows us to expand the homogeneous solution of the considered boundary value problem in a series in the eigenfunctions of the Laplace operator in rectangular coordinates and, thus, to obtain an exact solution.

2. Boundary value problem

We consider small fluctuations of the fluid in the areas τ_1 and τ_2. The fluid is assumed to be ideal and incompressible, its motion is potential, with the velocity potentials Φ_1 and Φ_2 in volumes τ_1 and τ_2 respectively. The compartments are separated from each other by a perfectly elastic, rigidly terminated plate, linear mass μ_0 and height H. Since plane oscillations of the system are considered, in the transverse direction the motion is translational [1], the plate is replaced by a homogeneous beam.

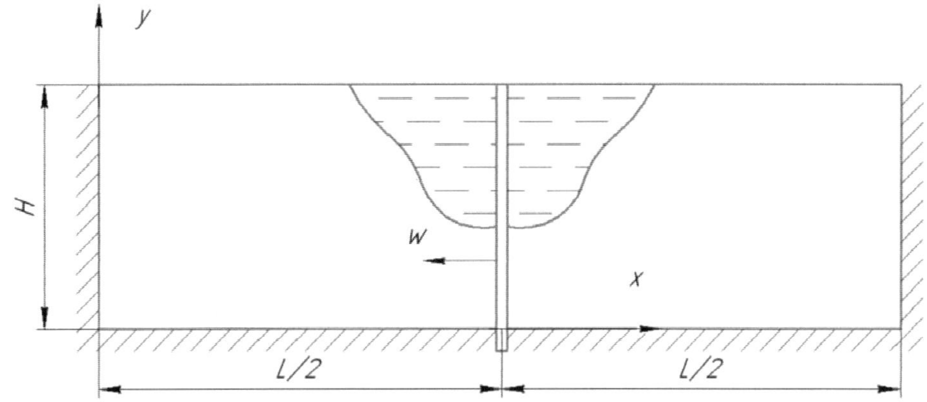

Figure. 1. Model of a tank with a liquid, separated by an elastic partition

The material of the beam obeys Hooke's law, the bending is straight, the bending stiffness of the beam is denoted as EJ_0 [2]. The dynamic pressure of the fluid through the Lagrange-Cauchy integral [8] is determined by the formula

$$P_i = \rho_i \frac{\partial \Phi_i}{\partial t} \quad i = 1,2 \tag{1}$$

Here ρ_i is the density of the liquid. In accordance with (1), the velocity of a liquid particle will be determined by the relation [8]:

$$\upsilon = -grad\Phi \ . \tag{2}$$

The velocity potentials are solutions of the Laplace equation [8] (Fig. 1)

$$\frac{\partial^2 \Phi_i}{\partial x_i^2} + \frac{\partial^2 \Phi_i}{\partial y_i^2} = 0 \tag{3}$$

According to the Fourier method [8], solutions for (3) have the form:

$$\Phi_1 = \sum_{j=1}^{\infty} \left(A_{1j} \cos \mu_j y + A_{2j} \sin \mu_j y\right)\left(E_{1j} ch\mu_j x + E_{2j} sh\mu_j x\right) S(t) \tag{4}$$

$$\Phi_2 = \sum_{j=1}^{\infty} \left(B_{1j} \cos \mu_j y + B_{2j} \sin \mu_j y\right)\left(D_{1j} ch\mu_j x + D_{2j} sh\mu_j x\right) S(t) \tag{5}$$

Here $S(t)$ is a temporary function; A_j, E_j, B_j, D_j are arbitrary constants to be determined [12, 7]. The differential equation of the bending of the beam [2] under the action of the hydrodynamic pressure of the fluid has the form [1, 2]

$$EJ_0 w_y^4 - \mu_0 \omega^2 w = \rho_1 \left.\frac{\partial \Phi_1}{\partial t}\right|_{x=0} - \rho_2 \left.\frac{\partial \Phi_2}{\partial t}\right|_{x_2=0} \tag{6}$$

Here, μ_0 is the linear mass, ω is the desired frequency of system oscillations, w is the deflection of the beam. The boundary conditions for the velocity potentials are:

$$\frac{\partial \Phi_1}{\partial x} = 0 \text{ on condition } x = 0, \quad -\frac{\partial \Phi_1}{\partial x} = \frac{\partial w}{\partial t} \text{ on condition } x = l_1$$
$$\frac{\partial \Phi_2}{\partial x_1} = 0 \text{ on condition } x_1 = l_2, \quad -\frac{\partial \Phi_2}{\partial x_1} = \frac{\partial w}{\partial t} \text{ on condition } x_1 = 0. \quad (7)$$

On the free surface of the liquid at $y = H$:

$$\Phi_1 = 0, \ \Phi_2 = 0. \quad (8)$$

Taking these conditions into account, the functions Φ_1 and Φ_2 have the form:

$$\Phi_1 = \sum_{j=1}^{\infty} A_{1j} \cos \mu_j y ch \dot{s}(t) \quad (9)$$

$$\Phi_2 = \sum_{j=1}^{\infty} B_{1j} \cos \mu_j y \left(ch \mu_j x_1 - th \mu_j l_2 sh \varpi_j x_1 \right) \dot{s}(t) \quad (10)$$

The eigenvalues of the boundary value problem are denoted by

$$\mu_j = \frac{(2j-1)\pi}{2H}, \quad \mu_1 = \frac{\pi}{2H}, \ldots \quad (11)$$

The equation for the plate (6) will be: taking into account (9) and (10)

$$\dot{w}^4 - B^4 \dot{w} = -g_1 \omega^2 \sum_{j=1}^{\infty} A_1 j \cos \mu_j y c g \mu_j l_1 + g_2 \omega^2 \sum_{j=1}^{\infty} B_{1j} \cos \mu_j y, \quad (12)$$

where $B^4 = {\mu_0 \omega^2}/{EJ_0}$.

The solution to equation (12) has the form:

$$\dot{w} = \dot{w}_{одн.} + \dot{w}^* \quad (13)$$

$$\dot{w}^* = \sum_{j=1}^{\infty} C_{1j} \cos \mu_j y + \sum_{j=1}^{\infty} C_{2j} \cos \mu_j y$$

Based on (7), we have:

$$C_{1j} = -\frac{g_1 \omega^2}{(\mu_j^4 - B^4)} (ch \mu_j l_1) A_{1j}, \quad C_{2j} = -\frac{g_2 \omega^2}{(\mu_j^4 - B^4)} B_{1j}, \quad (14)$$

Based on [2]:

$$\dot{w} = C_1 ch \beta y + C_2 sh \beta y + C_3 \cos \beta y + C_4 \sin \beta y + \sum_{j=1}^{\infty} A_{1j} \gamma_{4j} \cos \mu_j y, \quad (15)$$

where

$$\gamma_{4j} = \frac{\omega^2}{(\mu_j^4 - \beta^4)} \left(g_2 \frac{\cos \mu_j l_1}{\mu_j th \mu_j l_2} - (ch \mu_j l_1) g_1 \right) \quad (16)$$

Condition for fixing the console beam: $\dot{\omega} = 0$, $\frac{\partial \dot{\omega}}{\partial y} = 0$ on condition $y = 0$.

Taking into account the boundary conditions for securing the beam deflection will be:

$$\dot{\omega} = C_1 f_1(y) + C_2 f_2(y) + \sum_{j=1}^{\infty} \gamma_{4j} (\cos \mu_j y) A_{1j},$$

(17)

where

$$f_2(y) = ch\beta y +_1 \cos \beta y \,,\, f_2(y) = sh\beta y + \alpha_2 \cos \beta y \,, \tag{18}$$
$$\alpha_1 = \frac{ch\ \beta H}{\cos\ \beta H}\,,\ \alpha_2 = \frac{sh\beta H}{\cos \beta H} - tg\beta H\,.$$

Functions $f_1(y)$ and $f_2(y)$ we expand into series in functions $\cos \mu_j y$ [12]:

$$f_2(y) = \sum_{j=1}^{\infty} \beta_{1j} \cos \mu_j y\,,\quad f_2(y) = \sum_{j=1}^{\infty} \beta_{2j} \cos \mu_j y\,, \tag{19}$$

where

$$\beta_{1j} = \frac{\mu_j (-1)^{j+1} ch\beta H}{\left(\mu_j^2 + \beta^2\right)} + \frac{H}{2}\frac{ch\beta H}{\cos \beta H}\,,$$
$$\beta_{2j} = \frac{\alpha_2}{2}\left[\frac{\sin(\beta + \mu_j)H}{(\beta + \mu_j)} + \frac{\sin(\beta - \mu_j)H}{-\beta} - \mu_j\right] + \mu_j \frac{\left[1 + (-1)^{j+1} ch\beta H\right]}{\left(\mu_j^2 + \beta^2\right)}\,.$$

Based on (7), we define arbitrary constants:

$$A_{1j} = C_1 \beta_{1j}^* + C_2 \beta_{2j}^*\,,$$
$$\beta_{1j}^* = \beta_{1j}/\gamma_{sj}\,,\quad \beta_{2j}^* = \beta_{2j}/\gamma_{sj}\,, \tag{20}$$
$$\gamma_{sj} = \frac{\cos \mu_j l_1}{th \mu_j l_2} th \mu_j l_2 - \gamma_{4j}\,,$$

and the deflection of the beam looks like:

$$\dot{w} = C_1 f_1(y) + C_2 f_2(y) + C_1 \sum_{j=1}^{\infty} \beta_{1j}^* \gamma_{4j} \cos \mu_j y + C_2 \sum_{j=1}^{\infty} \beta_{2j}^* \gamma_{4j} \cos \mu_j y$$

Satisfying the conditions for fixing the beam, we obtain a system of homogeneous algebraic equations for determining the constants C_1 and C_2.

$$C_1 a_{11}\left(\omega^2\right) + C_2 a_{12}\left(\omega^2\right) = 0\,,\ C_1 a_{21}\left(\omega^2\right) + C_2 a_{22}\left(\omega^2\right) = 0\,,$$
$$\Delta = a_{11} a_{22} - a_{12} a_{21} = 0$$

The last expression is the frequency equation of the desired boundary value problem, where

$$C_1\left(f_1(0) + \sum \beta_{1j}^* \gamma_{4j}\right) + C_2\left(f_2(y) + \sum \beta_{2j}^* \gamma_{4j}\right) = 0$$
$$a_{11} = 1 + \frac{ch\beta H}{\cos \beta H} + \sum_{j=1} \beta_{1j}^* \gamma_{4j}\,,$$
$$a_{12} = \frac{sh\beta H}{\cos \beta H} - tg\beta H + \sum_{j=1}^{\infty} \beta_{2j}^*\,.$$

Provided $y = H$ we have

$$a_{21} = p^3 sh\beta H + \alpha_1 \beta^3 n\beta H - \sum_{j=1}^{\infty} \beta_{1j}^* \gamma_{4j} (-1)^{j+1}\,,$$
$$a_{22} = \beta^3 ch\beta H + \left(\frac{sh\beta H}{\beta H} - tg\beta H\right)\beta^3 \sin \beta H + \sum_{j=1}^{\infty} \beta_{2j}^* \gamma_{4j} (-1)^{j+1}\,.$$

IOP Conf. Series: Materials Science and Engineering 1191 (2021) 012024 doi:10.1088/1757-899X/1191/1/012024

3. Conclusions

In this work, according to the method proposed by P.K. Ishkov, an exact solution to the boundary value problem of plane vibrations of an elastically restrained plate is obtained. It should also be borne in mind that this method (the method of expanding a homogeneous solution in a series in terms of eigen solutions to the Laplace equation in rectangular, cylindrical spherical coordinates) was successfully applied in the works of P.S. Leibenzon, P.I. Balabukh and other scientists who dealt with hydroelasticity issues.

The solution obtained can be used to determine the dynamic characteristics of compartments in shipbuilding.

References

[1] Theoretical Mechanics Course. Ed. Acad. K.S. Kolesnikov. M., MSTU im. N.E. Bauman. 2017, p. 580
[2] Feodosiev V.I. Strength of materials. M., MSTU im. N.E. Bauman. 2016, p. 543
[3] Goncharov D.A., Pozhalostin A.A. 2018. *IOP Conf. Series: Journal of Physics: Conf. Series 012027*
[4] Goncharov D.A., Pozhalostin A.A. 2019. *IOP Conf. Series: Materials Science and Engineering: Conf. Series 012008*
[5] Pozhalostin A.A., Shtanko A.V., *Natural and technical sciences* **No. 3**, 2019, p. 19-22. DOI: 10.25633 / ETN. 2019, 03.04.
[6] Pozhalostin A.A. *Problems of mechanical engineering and machine* reliability. **No. 7**, 2019, p. 15-19.
[7] Leibenzon L.S. *Reports of the USSR Academy of Sciences*, 1951, **Vol. 1**, p. 125-133.
[8] Kochin N.E. and others. Theoretical hydromechanics. M., FN. GIFML. 1963.
[9] Kolesnikov K.S. Rocket dynamics. M., Mechanical Engineering, 2003.
[10] Balabukh L.I., Molchanov A.G.. *PMT*, **Vol. 6**, 1966, p. 1098-1102.
[11] S, Conlomb Theorie des machines simples (1821)
[12] Ishkov P.K. PMM. 1937, **Vol. 1**, pp. 1 - 18

AUTHOR INDEX